Ecodynamics of Insect Communities

Prof. T.N. Ananthakrishnan

SCIENTIFIC PUBLISHERS (INDIA)

P.O. BOX 91 JODHPUR (RAJ.)

Print : 2019

Published by:
Tanay Sharma
Scientific Publishers (India)
5-A, New Pali Road, P.O. Box 91
Jodhpur - 342 001

Tel.+ 91-291-2433323, 2624154
Fax.: +91-291-2613449

E-mail: info@scientificpub.com
www.scientificpub.com

ISBN: 978-81-7233-599-1

Laser typeset: Rajesh Ojha
Printed in India

Preface

My interest for well over fifty years on population based studies of mycophagous and gall inducing insects such as thrips, the degree of intraspecific diversity, the functional diversity of populations of several species and the degree of social life involved, has enabled me to think of such aggregations/communities in a host of other insects. The diverse definitions of what constitutes an assemblage, a guild and a community are basic to our understanding of their biotechnological aspects. Taxonomically restricted groups within a community comprise assemblages; guilds are species that share a common resource occurring in the same community, while all species occurring together in one place, form a community. Microhabitats such as dead and decaying wood, dung pats, plant galls, harbour diverse insect communities. All the same, large populations of a single species or species aggregations also need consideration. Community life is at it best among insect involving large colonies of ants, bees, wasps and termites. It is said that "social insects constitute the ecological centre, solitary insects at the periphery". Lawton (2000) has indicated that "most ecologists would probably feel comfortable using community for rather large assemblage of species, although there is no logical break in the continuity of a species with single species populations, to those involving hundreds of species." While it is almost impossible to study entire communities, an assessment of local assemblages to smaller communities would enable a better understanding of such group interactions. Lawton further comments that "Community ecology deals with patterns and processes involving groups of coexisting species at one locality and influencing one another directly or indirectly. Such plasticity allows for colonization and success in a novel habitat".

This book aims at an understanding of diverse mutualistic interactions, with all individuals tending to display their own adaptational ranges. In view of different species using different kinds of microhabitats, the diversity of assemblages goes a long way towards a better appreciation of the adaptive trends of the species involved therein. Insect assemblages and communities are shaped by the compositional, structural and trophic interactions which hold together ecological communities.

This book comprises 11 Chapters, the first three of which being essentially introductional, with aspects such as ecological diversity and biotic interactions; species aggregations and communities and nature of biotic interactions and the impact of habitat loss and fragmentation on insect communities. With the establishment of these three basic inputs, it becomes easier to appreciate the nature of microhabitats, their diversity, ecological and functional interactions as well as the ecology of mutualistic interactions, involving reciprocal benefits at individual/population levels, highlighting the diversity of such mutualistic associations such as that of ants-aphids, ants-coccids, ants-whiteflies and ants-membracids. Chapter IV deals with insect communities in natural and plantation forests and in agroecosystems. Emphasis is made on the wide range of responses among myriad species in natural forests to changes in forest quality. Similarly agroecosystems being ecological and socio-economic systems, the implications of such large populations of insect pest species are assessed. Chapter V deals with biological and ecological inputs in relation to insect sociality, involving ants, bees, wasps and termites. Chapter VI discusses the adaptive dynamics of insect-induced galling systems. Chapter VII deals with the ecology Invasive Insects. Chapter VIII outlines the diversity of interactions involved in insects infesting weeds. Chapter IX briefly outlines the functional aspects of Insect guilds. The last Chapter X deals with the climatic influence on Insects communities. This is followed by a short overview, followed by references.

In this attempt, I have received help from several friends who unhesitatingly provided me with photographs, which have made the book more usable. I wish to thank them individually to appreciate the significance of such assistance without which the book would not be presentable. My thanks are due to Prof. Raghavendra Gadagkar of the Centre of Ecological Sciences, Indian Institute of Sciences, Bangalore for the excellent photographs of diverse social Insects; to my son Ramdas A. Iyer of New Jersey for the photographs on migration of Monarch butterflies; to Dr. V.V. Ramamurthy, Principal Scientist, IARI, New Delhi for several photographs of Insect Pests; to Prasanth Jacob of the Centre for Tree Breeding and Forestry Research, Coimbatore, for the photographs on Eucalyptus galls; to Dr. Sabu Thomas of the Department of Zoology, St. Joseph's College, Devagiri,

Calicut, Dr. Alexander Jesudasan, Madras Christian College, Dr. A. Raman of the Charles Stuart University, Australia for their ready response to requests for reprints of publications and to Dr. G. Thirumalai of the Zoological Survey of India, Chennai for reproduction of several pictures from slides. To Dr. Ralph Holzenthal of the Entomology Department, University of Minnesota Minneapolis, my special thanks are due for the ready assistance rendered during my visits, and to the librarian for ready loan of books.

Last but not least, a special word of thanks to the publishers for their patient handling of the manuscript.

T.N. Ananthakrishnan

Contents

Chapter 1

Ecological Diversity and Biotic Interactions: An Introduction

Researches in biological diversity span through the themes of evolutionary biology, population and community ecology, besides the diverse multidisciplinary interactions. The essential ingredients of biodiversity pertain to phenotypic flexibility, community diversity, genetic variation within populations, ecotypic variation, besides species richness and diversity. Many are the indicators of ecological processing influencing community structure, which are essential for ecosystem stability and diversity. Everything revolves around the species, which are dynamic plastic elements changing their genetic constitution under the influence of the physical factors of the environment. Intensive studies on diverse aspects of insect-plant interactions have emphasized the role of a bewildering number of chemical inputs and the ability of insects to identify preferred hosts. This is astounding in that each species being endowed with unique sensory inputs enable recognition of its host range and adjusting optimally to changing conditions. Taking into account the structure and function of different communities, the responses of species to environmental changes appear individualistic and every community is different with different species and different webs of interactions in different environments. Taxonomically restricted groups within a community are referred to as `assemblages', while species that share a common resource and occur in the same community comprise `local guilds'. A taxonomically restricted group of species occurring in one place exploiting the same resource, such as gall-forming Diptera are called an `ensamble' (Lawton, 2000). In short, each and every community is different and all individuals display their own adaptational ranges. In different parts of their distributional range, species differ in abundance and the term `common occurrence' implies their adaptability to local environments. When populations constituting the communities are small, there tends be more diversity, because the coexistence of different species in a community is regulated by the basic resources available in the ecosystem. Sometimes one or more species may dominate when the term `Community dominance' is used. The structural and functional aspects related to such community structure, irrespective of whether they are macro or microhabitats, the impact of environmental degradation on insect communities inhabiting them appears considerable.

Adaptive flexibility of insects to respond to ecological and selection pressures is an important aspect. Their uncanny ability to choose feeding and oviposition sites naturally depend on their sensory capabilities, with information from the environment. Their endless diversity, as also of their habitats result in their encountering an extraordinary variety of seasonal and biotic conditions. Selection of suitable feeding and ovipositional sites with uncanny efficiency is typical of several phytophagous insects. The interaction of an insect pest with the components of the environment determines the size of the population. Such interactions may often be mutualistic as those between pests associated with the crop and parasites and predators; between all pests inhabiting the crops; between the insects pests and the food chain; between the insect, it host plant and environmental conditions and those between the insect species and environment stimuli. The impact of an insect on the environment can best be understood not by the presence of all factors, but by the absence of any one of them which would produce visible effects on the individuals of a biotic community.

Ecologically insects perform a striking variety of functions, being involved in such critical ecological processes like plant protection, pollination, species mutualism and competition, ecosystem maintenance and stability, decomposition, predator and parasitoids. Certain insects, the `keystone species' play central roles in maintaining entire biological communities. Species diversity in tropical forests results from varied interactions and interdependency among plants and insects, notably such roles of insects as pollination, seed dispersal and plant protection. It has to be emphasised that communities change in response to basic conditions of the environment which tend to affect the insect fauna. All communities at the end or the beginning of succession are drastically different, the degree of diversity increasing as succession proceeds. Some species modify their physical environment and hence the environment of other species, often with profound consequences and such modifications have been termed `ecological engineering' by Lawton (2001).

The synthesis of ecological, behavioural and evolutionary aspects of populations go a long way towards a better appreciation of community biodiversity. For example, insect parasitoids being an integral part of the complex of multispecies interactions, an understanding of the taxonomic and functional aspects of community structure become relevant. Taxonomically different, but functionally different species, often referred to as `species guilds' flourish at great heights in the canopy of forests, harbouring such guilds in countless numbers, noted for their ecological services in the form of pollinators and biocontrol agents. The need for an assessment of how species respond to environments and how evolution functions in species abundance of insects which show considerable intraspecific diversity. Species being dynamic, plastic elements, population studies establishing variation patterns appear important and basic to an understanding of taxa. An assessment of single species polymorphic aggregations becomes significant, as they occur, with a high degree of intraspecific diversity, also involving functional diversities, very typical of several species of mycophagous and gall forming thrips, coccids and aphids. As such populations based on ecological studies enable identification of polymorphic populations. The complexity of ways in which species specialize and evolve, the pattern of specialization and adaptive diversity are aspects which have engaged the attention of biologists in recent years. Approaches to such specialization have developed from ecological, physiological and biochemical studies notably from insect-plant interactions. The potential for interacting with closely allied species tends to create different degrees of specialization, not to emphasise the fact that specialization in one environment, becomes opportunities for new interactions in other environments, so that specialists evolve from generalists. Variation being an important aspect of evolution, appreciation of aspects such as behavioural patterns, adaptation to climatic variations and the ecological conditions become relevant. For instance, certain insects must aggregate on host plants in order to survive, which often depend on an even distribution of populations to prevent overcrowding in areas of limited resources. A example relates to pine trees with the aggregation of bark beetles, mediated by pheromones produced by the invading beetles. On the contrary, often a host plant will be able to support high densities of one species without reduction in yield, but much lower densities of a species tend to cause economic damage.

Behavioural diversity within a population is common, not to mention of molecular markers which have greatly enhanced thoughts of intrapopulation variation. Within populations, genetic variation become useful parameters to determine the potential for rapid evolutionary change, enabling insects to adapt to rapid changes. Within a population, groups of individuals exist, that utilise different hosts and these are genetic variants, besides species which show population differences in host use patterns. Several species of butterflies use different host plants in different regions so that their population becomes geographically isolated with increase in genetic differences (Bernays-Chapman, 1994). Interestingly enough, there exist strains of insects with inherited differences in their ability to use host species. For instance, whenever, a crop variety is no longer resistant, new `biotypes arise which are polygenic variations within populations. Recognition of such structural and functional plasticities becomes essential. The potential for closely interacting species tends to create different degree of specialization among populations and species, not to emphasise the fact that specialization in one environment creates opportunities for new interactions in other environments.

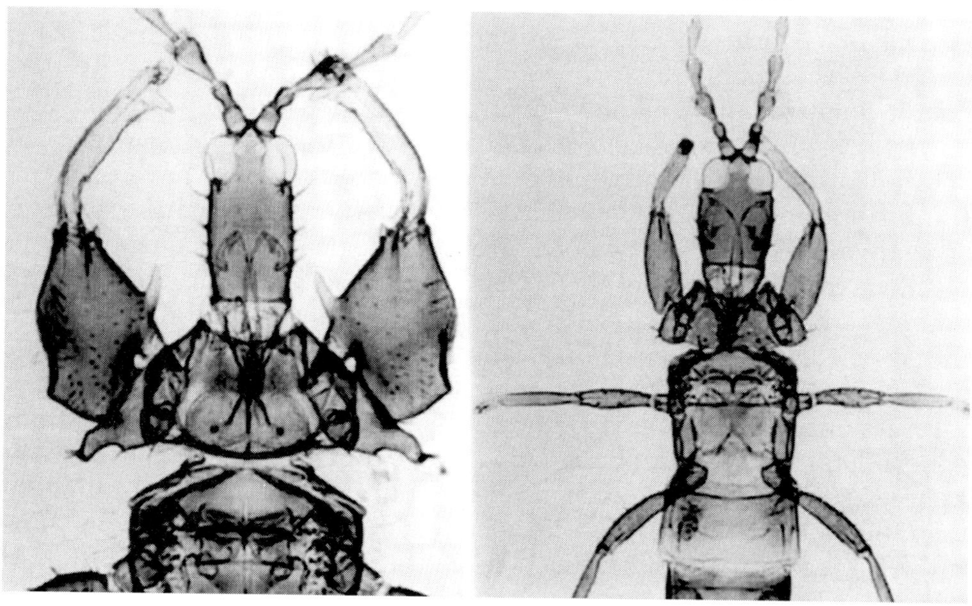

Fig. 1. Phenotypic plasticity in *Ecacanthothrips;* Left, Oedymerous male, right; Gynaecoid female.

Seasonal dimorphism as in many butterflies, aphids, leaf hoppers are typical instances of adaptive flexibility. The impact of individual variation on the reproductive capacity of an insect or the genetic release mechanisms involved in the production of diverse phenotypes appear important in the reproductive biology of a species. Such a knowledge becomes essential for an appreciation of structural, physiological and developmental strategies enabling prolonged survival of individuals and the formation of superior mating preferences. This is well exemplified in studies relating to the superior fitness of morphs with respect to reproductive efficiency and with increased number of individuals being produced, polymorphic individuals become better suited for breeding and population build-up. Ecological strategies of a species are evolved in response to the habitat and the `reproductive success matrix' implies that individuals of a species, in time and space tend to evolve, strategies which would maximize the number of their descendents to the habitat (Southwood, 1977). Further, the distribution and abundance of a species in a multitrophic community in influenced by phenotypic plasticity.

What is phenotypic plasticity?

The ability of an organism to express different phenotypes depending on the biotic or abiotic environment, often referred to as *phenotypic plasticity*, plays an important role in the adaptation of insects, more especially in relation to their ability to thrive in changing environments. To be more specific, phenotypic plasticity is the expression of multiple phenotypes by one genotype allowing individuals the flexibility to alter their morphology, physiology and behaviour (Stauffer and Whitman, 1997). In other words polymorphism contributes considerably to our understanding of the casual factors underlying variation. Besides, morphological specialization, is the broad tolerance of individuals in a variety of ecological conditions (*broad phenotypic plasticity*) as against that for a narrow niche. The existence of genetic polymorphism within species, producing individuals to a variety of ecologically adapted sub-niches is well known, besides the occurrence of ecotypic variation among populations within the geographic areas of species (Thompson, 1994). In short, groups of individuals within species have the potential to become specialized to exploit different habitats. A good example is the genetically identical female morphs as seen in some aphids showing differences in host plant preference, morphology size, number and timing of reproduction of offspring. The consequences of phenotypic plasticity plays an important role leading to ecological and evolutionary success of organisms. Besides morphological aspects, such phenotypic variation can also occur in life history traits, size clutches, larval growth rates, reproduction and survival. Richards (1961) defines polymorphism as "one or more sexes of a species occurring in two or more forms which are sufficiently sharply distinct to be recognizable without monorphometric analysis. The occurrence is regular, the rare of the two forms makes up a reasonable proposition, the rarest type being essential for the survival of the species". Hence environments experienced by a population determine the phenotypes and establish their fitness for a particular trait. Phenotypic plasticity has to be regarded as a fundamental character of insects, subject to evolutionary pressure, not merely an intrinsic variation an aspect deserving increased attention by those involved in studies on biological diversity wherein phenotypic plasticity is an integral component.

The response of species to environmental changes appear individualistic and every community is different with different species and different of interactions. Taxonomically restricted groups within a community are referred to as assemblages. While species share a common resource and occur in the same community, comprise local guilds, while taxonomically restricted groups of species occurring in one place exploiting the same resources, such as gall forming Diptera are called an *ensemble* (Lawton, 2000). In short, each and every community in different and all individuals display their own adaptational ranges. In different parts of their distributional ranges, species differ in abundance. The term phenotypic plasticity refers not only as a relative immediate response to an environmental change. Bernays and Lewis (1980) showed that caterpillar head size responded plastically to the toughness apparently providing the appropriateness of musculature for individuals feeding on harder leaves. Similarly, Thompson (1992) demonstrated that plant induced changes in mandibular morphology increased feeding efficiency. Such plasticity acts as a mechanism that responds to natural selection and allows for colonization and success in a novel habitat (Mooney & Agrawal (2008). To quote Whitman and Agarwal (2009), the greatest limit of phenotypic plasticity is the way it generates adaptive change, an essential long term strategy for persistence and their adaptability to local environment. Ecosystem, communities, species and populations lack clean boundaries in space and time and interspecific interactions may go a long way in promoting a better understanding of species interactions. Keeping these aspects in mind, emphasis is laid on the diversity of insect communities, besides the functional role of diverse species in such interactions. It may often occur that a species or more than one species may be relatively abundant among the various species populations in a community and such species may exert a control over the other inhabiting species. Such dominance is conspicuous in habitats with special environmental conditions, as in deserts. In short, the dynamics of populations and the degree of persistence of species become important, not to mention that small fragments can harbour only smaller populations, which tend to have genetic variation – a warning toward risk of extinction. It is this migration among fragments become important, since it is only through migration that fresh populations tend to replace those threatened by extinction (Hanski, 2005). Phenotypic plasticity in a novel habitat may lead ultimately to evolutionary divergence and specialization. Insects tend to use greater diversity of phenotypic plasticity than plants. This is because of the intricate interactions involved in trophic diversity, such as tritrophic interactions. The publications of Ananthakrishnan and Whitman (2005) and Whitman and Ananthakrishnan (2009) stand further testimony to the significance of phenotypic plasticity in insects.

Chapter II

Insect Communities and nature of species Interactions : An Analysis

Community ecology in general, tends to focus on aspects relating to direct and indirect interaction within and between species, enabling a better appreciation of their structural and functional components. While communities are groups of species that interact or have the potential to interact with one another, they comprise taxonomically related groups including species richness, dominance, spatial arrangement and length and organization of food webs (Lawton, 2005). The ever changing composition of biological communities tends to create opportunities for newer interactions. As species are introduced into new areas, they are stripped of their earlier interactions and acquire new ones (Thompson, 1991). As populations evolve they form new interactions and specialize in new directions. Individual interactions always tend to cause changes in population, not to overlook the consequences of environmental change on intraspecific diversity. With environmental change or variability, populations also tend to change, with diversity in relation to the population fluctuation patterns. In short, this structural relationship between insect aggregation/ communities and their living environment is well reflected in the concept of the ecosystem which is a functional entity comprising a community of living organisms and the physical environment in which they live, enabling closer interactions. Ecosystems such as the forest ecosystem comprise a community of living organisms of diverse kinds with groups of individual species of insects or of several species discharging a similar function. To be more successful, environmental variability on the part of insects appear to be an essential aspect, aiding the development of structural, physiological, genetic and behavioural characteristics which enable them to behave differently in different habitats, at different periods of their development and under varying conditions of the environment.

The everchanging composition of biological communities has resulted in opportunities for new interactions between species. The complexity of ways in which species specialize and evolve, the patterns of specialization and adaptive diversity are aspects engaging the attention of biologist in recent years.

Intraspecific patterns tend to be as much important for community ecology and as Lawton (2005) opines, "species play geographically variable roles in their communities, that species differ in abundance in different parts of the grographic range and common species are not common everywhere". Further, the impact of a species on its community at one locality, with not often be the the same as its impact as on another locality Hanski (2006) classifies four types of habitat loss loss of quantity, quality, of connectivity and loss of connectivity. Habitat fragmentation increases with decreasing sizes and increasing isolation of species of the habitats with which a given area has been split. Habitat fragmentation tends to reduce the rates of migration as also the rates of immigration and gene flow among populations in the remaining fragments.

While species diversity and complexity of association among species are essential for the stability of a community, an understanding of the components of diversity tends to the useful in assessing community diversity. Basically they involve (a) the number of species and their relative importance (b) intra-and interhabitat diversity (c) diversity of resources, average utilization and overlap (Pianka, 1978). While the number of species in a sample is easy to determine, a natural community is an open system without a fixed number of species and hence the usage of term species richness. Greater number of effective habitats support more species as in the case of the forest ecosystem, so that the turn over in species composition varies with more effective habitats – defined as within habitat diversity. There is alway a certain degree of overlapping or intergrading of habitats with one another, typical of ecotones bordering forest habitats or with agroecosystems bordered by weeds. Massive invasion rates also result in large scale overcrowding of single species populations, such as that of *Mallothrips indicus* on *Syzygium jambolanum* with thousands of larvae and adults on the tree trunk, fallen leaves and fruits. The same is true of *Isoneurothrips australis* on a single eucalyptus tree in Australia (Laughlin, 1970), as also that of *Parabaliothrips* within the terminal bud of *Ficus microphylla* with male aggregates of 80-110 (Gillespie, 2002).

The diversity of resources available in a community determine the variety of opportunities for ecological diversification within it.

Communities with restricted resources will naturally support lesser species those with greater resources. However, as the degree of utilization of resources increases, the number of species that can coexit in a community decreases. A community with a greater overlap will support more species than one with less overlap. This is very true of insects on crops particularly along their marginal zones where wild vegetation exists. While habitat in the most fundamental necessity for the survival of populations, the locality with its particular type of environment and the resources and condition present in the area enable survival and reproduction of a given species. Whittaker (1973) opined that the term habitat refers to the "range of environments over which a species occurs, whereas *niche* designates the functional role of the species in the community. Populations of one species at many sites are referred to as *metapopulations* or an assemblage of local populations inhabiting a network of habitat patches. The diversity of species and their interaction together form biological communities which are part of the very fabric of what makes a species and a biological community" (Thompson, 1994).

If the total habitat encompasses a number of describe grouping of individuals interconnected by some limited degree of dispersal, metapopulation result. For example, populations of the coccinellid *Ephilachna* are subdivided and connected by adult movement, limited to a small range. The number in the subpopulation tend to the stabilized through interchange among them, diversity dependent adult mortality and larval competition for bond (Hanski, 1997).

The everchanging composition of biological communities tends to create opportunities for new interactions. An understanding of species and their interaction is basic to an appreciation of the structural organization of communities and in this organization, the role of competition becomes important. In short, the study of evolutionary interactions demands the mixing of so many subdisciplines, ecology, systematics and genetics. Ecological impact in relation to say, Ant-Hompteran association is an aspect which cannot be overlooked. Large numbers of aleyrodids, coccids, aphids and membracids form interesting association with ants, which together form an "ecological novelty", in the sense that mutualism plays an important role in the maintainance of aggregates of different species. Similarly the term *biotope* is used for the entire community of many species, such as that of dung beetles belonging to the cattle pasture environment. Hanski and Kosaba (1977) in their paper on `Niche relations among dung inhabiting beetles, indicate that 62500 beetles were extracted from hundreds of cowpats, with 179 species of beetles".

In the assessment of ecological succession, the three fundamental aspects relate to, impact of resources on populations of species succession/dominance and the impact of natural enemies on population abundance. Alteration of ecosystems through human manipulation, have led to structural and functional differences in natural ecosystems which have high species diversity, genetic diversity and stability. Existing communities are disrupted by deforestation and agricultural practices resulting in new man-made communities, with aspects such as insects-weed colonization. While natural selection and coevolution result in diverse assemblage of insects, diversity is sufficiently reduced by human management practices. But changes in landscape diversity have resulted in insect outbreaks through decreasing habitat diversity, often affecting both diversity and abundance of natural enemies.

An understanding of insect community patterns and processes become relevant, with parasitoids playing a significant role in the assessment of species richness. Various species are organized into one or more guilds which are groups of species which exploit the same environmental resources in a similar way. Guilds are often considered as building blocks of ecological communities and play an important part in community ecology. For example, the parasitoid guild is an ecological unit, not a taxonomic one. As a consequence of the large size of plants, insects can evolve to feed on a single plant in diverse ways, such as by feeding caterpillar, others mining through the leaf, others like aphids, leaf hoppers, bugs sucking sap, cynipid wasps inducing and feeding on galls, weevils chewing the embryo in the seeds, so that plant complexity is an important determinant of the number of species that can few on a plant. Besides the communities, the spatial occurrence of dead wood microhabitat with the available resources in sufficient quantity and the important role of breeding insects, with large logs tending to support populations of small insects constitute an important area of community ecology (Hanski, 2005). Other microhabitats providing adequate resources are the tree holes, dung pats, pockets of soil in tree canopies, which do not persist for a long be time, the species involved tending to be good dispersers. Butterflies for example, depend on nectar resources of flowers and their survival is enhanced by this abundance which influence their movement patterns, with different species differing in terms of availability of resources for reproduction and survival. The habitats of Monarch butterflies have been threatend by extensive logging. By removing a single tree, heat escapes from the frost, jeopardising the butterflies. Monarch butterflies have become an ecological symbol, their migrations are proverbial. Each September they begin a 5470 km

Journey from the forests of. East Canada and parts of the USA to the Mexican central mountains. In the words of Lawton (2005) `every community is different with different species and different webs of interactions in different environments. Interactions within one community at one place and time from place to place tend to decay and change over time from place to place, not to mention of the individualistic nature of response to environmental changes".

Fig. 2.1. Monarchs resting in clusters (after Ramdas Iyer).

Communities differ with different species and webs of interactions in different environments. In view of the large scale literature available and indepth discussion provided by Lawton (2005). Efforts are made here to involve several instances of aggregations of individual species and their intraspecies diversity. Interspecific competition for resources is typical of communities with populations varying with variation in the environment over time. Various patterns of population fluctuation are evident in different species, besides the ecological succession of species inhabiting forests, agroecosystems, ecotones, as well as rare instances of succession relating to different species, such as that of thrips occupying different nodes on the same plant. As environments change with time both absolute and relative abundance tend to change (Lawton, 2005). Further, local communities are not hermetically sealed units; species enter and leave them, so that local interactions and processes are also subject to large scale influences", so well illustrated by Lawton (2005) in his studies on bracken fern insects. Further, the shape and spatial arrangement of habitat patches tend to affects population, persistence and dynamics". Nair (1998) has shown that the foliage of about 7260 hectare of plantations of teak was almost uniformly consumed by the outbreak of populations of *Hyblaea peura.* Similarly ants account for 20-40% of the arthropod biomass of the canopy.

Fig. 2.2. Aerial migration of monarch butterflies (after Ramdas Iyer)

Each species has a specific position in a community referred to as the *niche.* Interestingly enough, insects totally unrelated to each other, yet occupying identical niches in widely seperated areas, is a common feature. This is due to similar environments enabling an identical community structure. Evolution of species in a community is determined by the structure of the community, which in turn is dependent on the abiotic environment. When populations constituting a community are small, there tends to be more diversity and the coexistence of several species in a community is regulated by the basic resources available in this the ecosystem. The transition zone between two communities is the *ecotone* which present conditions that are intermediate to communities on either of it. Ecotones present a perplexing diversity of species due to the occupation of the transition zone by the various species of the major communities to ordinary the ecotone.

Fig. 2.3. Monarchs on the wing (after Ramdas Iyer)

The tendency for communities to change is common and the replacement of one by other is called ecological succession, which though gradual is a continuous process in which one community gives place to another, the degree of diversity increasing as succession proceeds. A good example in that of a dying or a rotting tree. The first to occur are the fungi, and beetles helps to loosen the bark, which makes room for other species of insects, followed by predators, the last to appear being the termites. The earlier stages and the pioneer communities, the final ones being called climax communities. Biotic factors of the environment also tend to modify the activities of insects. Temporary or permanent aggregations hold the animals as a unit as in the case of the membracid bugs. Scarcity of space, utilization of the same food and shelter and collective breeding induce them to aggregate. On the other hand carrion or dung, fungal fruiting bodies, fallen logs or decaying wood debris are good microhabitats. The more persistant microhabitats are dominated by beetles. Dead wood microhabitats have played a special role in the evolution of insects, since they involve wing polymorphism, sex dimorphism, male haploidy and social life as in termites and outs. Gall inhabiting species occurring in one place exploiting the same resource are called an *ensemble.*

The impact of a species on a community at one locality will not often be the same as it impact at another locality. "For a majority of species, richness in assemblages and local guilds appear to be primarily determined by the changes in size of the regional species pod and the role that any one species plays within a community varies spatially within its geographic range. A major challege to ecology is to understand more clearly the role of regional processes in local community dynamics" (Lawton 2005).

Ecological succession

Communities change in responses to the change in the basic conditions of the environment. The phenomenon of the replacement of one community by other members of a community in course of time is called *ecological succession*. It is a gradual and a continuous process in which one community gives place to another, a process of self organization based on interaction among species and between the organisms and the environment. Ecological succession is *primary* when it occurs in a habitat not previously occupied by any community; it is *secondary* when it occurs when the inhabitated communities are removed. While primary succession is a long term phenomenon, secondary succession occurs more rapidly. The first community of organisms to become established are the pioneer community, which gives way to other with the laper of time. A stage occur when succession reaches its zenith so that the community at that time is in perfect equilibrium with the environment. This final community is the *climax community*.

In successional ecosystems open habitats are the target of colonization with modification of niches and this results in competition. As indicated earlier, the succession of small subunits of a community also occurs as in the case of a dying or rotting tree, and such an orderly dying true of small subunits of a community. Establishment of a climax community is never a finale to the fascinating changes occuring in an ecosystem. An established climax community however much in dynamic equilibrium with the environment still starts evolving.

Southwood (1977, 1988) has developed a theme relating to ``habitat as the template", that shape the ecology, and evolution of species. Plants provide the template at the base of the for chain, with a bottom-up perspective having considerable impast on upper trophic lerch. Insects respond to this habitat template generally with added complexity for parasitoids (Price, 1994). Needless to emphasise that plant hosts and all kinds of vegetation, with their structural complexity, increases habitats in all direction. As plant community changes insect herbivores on hosts also change. As the parasitoid community develops on a host, species richness increases with increasing competition with ectoparasites generally overcoming endoparasites.

Habitat loss and impact on insect communities

Natural ecosystems are endowed with high species and genetic diversity, with heterogenous habitats endowed with species rich insect and plant communities. Increased resource exploitation and alteration of land use patterns have resulted in the fragmentation of habitats, ecosystem and landscapes. Fragmentation increases local rate of extinction by reducing population size or colonisation from similar habitats and by changing the physical and biological environments. Survival of species is invariably affected by fragmentation of ecosystem and of biotic communities, of species population and habitat range of a particular species. Nair (1991) defines forest fragmentation as the seperation of forest areas by non-forested areas, where the land use practices are opposed to natural re-establishment of forest vegetation. The sequential process of ecodegradation involves loss of biodiversity, stability and productivity of the forest ecosystem, with fragmentation of extensive forests into smaller segments. In general, habitat loss involves loss of quality and loss of microhabitats such as that of dead and decaying wood. The net result is that degradation, fragmentation, logging and farming tend to intensify agriculture, with the overall loss of microhabitats (Hanski, 2005). Different species have different ecological requirement making habitat deterioration vulnerable to species loss or reduction. In short, habitat loss and fragmentation tend to change the phenotypic and genotypic structure of populations, reducing their migrating rate. Intensive studies on the density and habitat occupancy of species could indicate the possibility or otherwise of a species to become extinct. For instance, while the density of the coarse decaying wood is high in natural forests, the occurrence of the microhabitat speciality species is equally high; the loss of such microhabitats tends to lead to species loss due to the effects of habitat loss and the degree of fragmentation.

Insects as indicators of land use change

The type of land use change will tend to affect the diversity of species composition of insects such as ground beetles and moths involving variation in intensity. Soil disturbance tends to have a severe effect on larger species of carabids, soil moisture especially being a major detriental fealm in beetle assemblages. Needles to emphasize that overall insect diversity in the Tropics is substantially affected by landuse management both in agriculture and forestry.

Such effects could be studied at the level of the individual species, on the composition of species in the habitats, species richness and diversity. Ants, lepidoptera larvae and carabids are often used as indicators of land use change. Needless to emphasize that fragmentation of habitats into smaller, isolated patches present a severe threat to poorly dispersing species.

Considerable changes tend to arise in species complement, diversity, numbers and biomass of insects. Local extinction or survival of several species as a consequential change in land cover tend to arise. For a better assessment of such change the need for application of ecological concepts to the strategies of individual species becomes relevant. Further, the very existence of a particular species depends on the pressure of some other species and the extinction of one species may lead to the extinction of one or more species resulting in a chain of extinctions. Brook *et al.* (2003) estimated that Singapore lost 40% of its butterflies in response to 95% loss of it forest habitat in 183 years. In general, extinction of butterflies is related to the extinction of their host plants. Locally speaking, the once lush Tenmalai forest of the Western Ghats had a complement of mycophagous and gall thrips. Within a course of 25 years more than half the species have disappeared through deforestration and replacement by plantation crops. In short, the disturbed landscape tends to cause perturbation leading to a chain of events leading to the extinction of several species of insects. Patches of rare insect communities are more susceptible to disturbance such as population fragmentation, inbreeding depression, increasing loss of genetic variation etc. Large tracts of forest which may the fragmented, result in increased number of edges, having implications for being susceptible to increased distribution of weevils and often plant species and associated insects. Strip corridors have an edge effect on either side and this narrow overlap zone between adjacent plant communities are highly significant to abrupt changes in species composition. Ecotones as they are called are highly significant for insects whose lifestyles are linked to the habitat. Needless to emphasize that diverse taxonomic groups are useful in the identification of particular landscape or of communities. In short, insects with a large variety of species and populations tend to act as environmental monitors and enable a better understanding of the impact of the landscape as good indicators (Samways 1994).

Chapter III

Microhabitats : Community diversity and mutualistic interactions

It is well known that many macrohabitats tend to become increasingly fragmented resulting in increased opportunities for diverse species to ocupy much smaller areas, having shorter life times often due to the action of decomposers that colonize them. Most microhabitats, though covering only a tiny portion of the total landscape area, are highly dynamic, with old patches constantly disappearing and new ones appearing, posing the greatest threat to a very large number of species. Most of the inhabitants are small-bodied insects, usally dependent on the concerned microhabitants. Increasing loss of such microhabitats such as fallen decaying logs, fungal fruiting bodies, dung pats of ruminants and elephants, corpses of decaying animals, living and dead plant galls have adverse effects on biodiversity. They also provide useful model systems for studies on the dynamics of species inhabiting them. Herbivore dung and carrion, rich in nutriants are colonized by insect communities whose behavioural, physiological and ecological aspects are among the most competitive of all animal communities (Hanski, 2001). The rapidity with which they colonize increases the rate of resource exploitation. Such studies have led to the development of *`aggregation models of competetive existence* (Hanski, 1981, 2001) enabling the species to coexist. Such intraspecific aggregations are very common in dung and carrion microhabitats. Similarly saproxylic species specialising on dead wood are equally diverse, but the great reduction of such habitats has adverse consequence for many saproxylic species.

Water-filled tree holes are semipermanent habitats susceptible to drying up and they are discrete habitats with the ability to resist extremes. Organisms such as mosquitos, chironomids, biting midges and mould better are common. These species are broadly divisible with accidental inhabitants, attracted to semiaqutic habitats. Similarly there are arboreal ants which are site-specific in the canopies of forests, exhibit vertical stratification, with considerable trophic specialisation such harvesting exudates, scavenging insects and capturing live pey. Besides these microhabitats the litter microhabitat in Tropical forests host a wide variety of insects many of which herbour a wide variety of species enabling a better insight into species population dynamics, not to mention of several mutualistic interactions evident ant homopteran insects, As microhabitats such as these do not persid for a long period of time, the need for an assessment of species dependent on them, becomes an useul exercise.

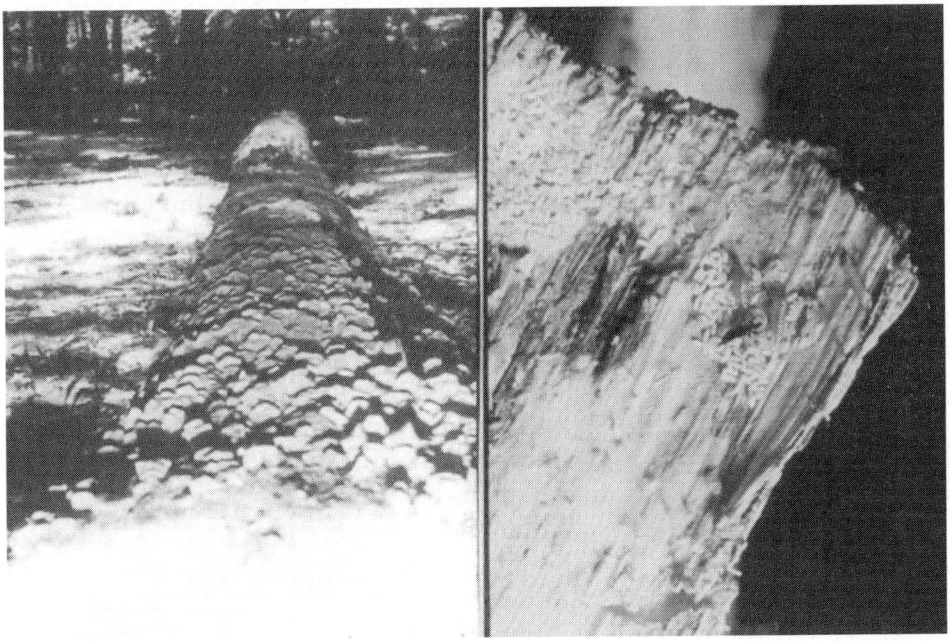

Fig. 3.1. Fungal infestation in dead wood, right: dead wood and a thrips colony

Wood inhabiting fungi and insect colonists

The fruiting bodies of macrofungi provide shelter and food for several insect species. While woody fungi harbour a host of insects such as collembolans, thrips, beetles, dipteran larvae, short-lived fleshy fungi are inhabitated by a large number of fly larvae including those of drosophilids, feeding on the fruit bodies which persuit for 2-3 weeks. Obligate fungus feeders spending their entire lives or atleast part of their life cycle, occur in the fruiting bodies of some fungi. Some polyporoids or bracket fungi are known to harbour more than 250 species of insects. *Fomes fomentarius*, a polyporoid wood inhabiting fungus causes serious diseases on standing and structural timber.

Succession occurs in relation to over 150 species of insects, the sporophore community comprising major mycophagous, predators, parasites, scavengers and tunnelers such as nitidulid and staphylinid beetles (Mathewman and Pileou, 1971). Of the dipteran families of primary fungivores feeding on the basidioscarps of fleshy fungi are the Phoridae, Mycetophilidae and Anthophoridae, the secondary fungivores being the Drosophilidae and Tipulidae. Cecidomyids have been known to be secondary fungivores feeding on decaying basidioscarps. Siricid wood wasps feed on the fungus *Amylostereum,* while the macrotermes groups of termites culture the fungus *Termitomyces.* Ambrosia beetles including the platypodids and scolytids culture different species of fungi in the pockets of their prothorax. Fungus feeding ants culture fungi on freshly ant leaves and plant debris, while cecidomyiidae galls harbour fungal mycelia, constituting typical examples of insect-fungus interaction, not to overlook the diverse mycetophagous and sporophagous thrips in dead tree habitats, some of which are indicated below:

Intraspecific diversity enables a combination of resources scattered in space and time to provide a composite niche for a species, enabling is to put forth a number of evolutionary opportunities - an aspect well exemphified by mycophagous Tubuliferan Thysanoptera. Mycophagous thrips is widespread, many species inhabiting a variety of microhabitats generally offered by dead trees which offer excellent breeding sites. Habitats underlying bark are extremely productive examples of wing polymorphism, of male haploidy, special breeding structures due to social or subsocial behaviour and of the development of certain types of sex-limited diversity, very typical of mycophagous Tubulifera (Hamilton, 1978). A further feature of dead tree complexes is that many species tend to readily mate within this habitat, improving the potential for increase and colonisation, contributing to the development of subsocial behaviour. Connected with this subsocial behaviour is the development of numerous sex-limited morphs, notably that of males.

These dead tree habitats exemplify productive habitats enabling gregariousness resulting in more effective fecundity and consequent increase in population size. Variations are magnified to a large extent by a high degree of heterogony or absolute size allometry of adults, making the morphological difinition of species more difficult. A finely integrated series is often noticeable in the population of such species and the two extremes are often referred to an the major or oedymerous males and gynaecoid or minor males. Oedymerism involves the development of bizarre forms with strikingly enlarged parts affecting mostly the structure and armature of the forelegs and incidental changes in thoracic structure. Gynaecoidism results in opposite traits, with weakly developed forelegs with highly reduced armature or even without it, resembling the females in general make up. The patterns of diversity in relation to such armature vary with species. When the form and abundance are known, it enables statistical interpretation of the growth rates of various parts of the body in different morphs.

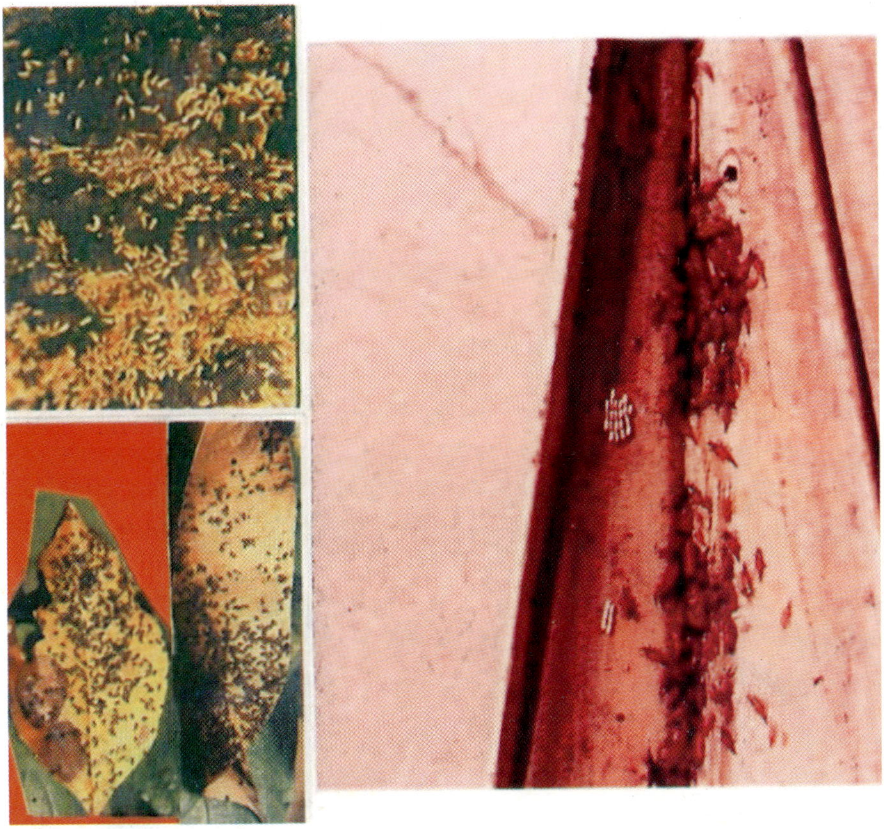

Fig. 3.2. Thrips in fungus infested leaves.

Extensive development of structures associated with oedymerism is strongly suggestive of male-male conflict in sexual selection. Crespi (1985, 1986) has indicated that the `island habitat' structure as evident under bark, with large scale inbreeding, accompanying gregariousness, overlap of generations with females living much longer than males, haplodiploidy, sperm storage by females and maternal care of eggs contribute to parasocial conditions. Excessive development of structure associated with oedymerism is strongly suggestive of male-male conflict in sexual selection. Such diversity is reflected in a host of species and Ananthakrishnan has analysed the patterns of ocdymerism in a host of mycophagous species. New features tend to express themselves in the normal - oedymerous transition and some of the notable examples are very evident in *Ecacanthothrips, Dinothrips, Tiarothrips, Bactridothrips, Cervothrips,* to mention a few.

The armature of the cheeks are equally noticeable. Instances wherein such morphisms are evident in the females are evident in *Machatothrips, Arrhenothrips, Mesothrips, Mallothrips* where the extreme females are called major females and /minor females. Major females have enlarged forefemora and foretarsal teeth and foretibial tuberclu in the major females. Numerous instances are known in a host of species and territoriality and fighting in species of *Hoplothrips pedicularis* living in colonies on fungi are evident in several other species. The structural, physiological and developmental strategies enable prolonged survival of invidividuals, promotion of superior mating preferences and in species where parasociality exists, care of oppspring. With increased number of individuals being produced, polymorphic individuals are better adapted to exploit the environment, being better buffered against environmental variations. The identification of functional groups involving species that have similar interactions, the effects of diversity among such groups depend on the degree of intraspecific diversity. Speciation and evolving novelty run rampant involving inbreeding in respect of their living in spaces offered by dead trees.

Fig. 3.3. Dung rollers *Onthophagus* sp. in Elephant dung (Courtesy, Sabu Thomas)

Fig. 3.4. dung roller beetles (Courtesy Sabu Thomas)

Dung beetles : guild structure and succession

Dung beetle communities are strongly influenced by dung type, many being generalists, while very few are specialists. Some dung beetles tend to select fibred dung of non-ruminants, while others prefer the more fluid and finer dung of ruminants. Two types of dung beetle communities relate to those associated with larger dung pats (eg. elephants and gaur) and smaller droppings of deer and boar. Many species occur in large populations and resource partioning and competition are prime factors of dung beetle species assemblages (Hanski, 1991). Three functional groups are known among during beetles, based on their nerving strategies - *rollers, tunnelers and dwellers,* with rollers forming balls from during parts, tunnelers creating underground chambers, while dwellers breed in the dung pats itself (Sabu Thomas *et al.,* 2006). This functional stratification allows dung beetles to minimize the intense competiton for limited food and space. Resource competition between the rollers, tunnelers and dwellers is very evident as entire pats/droppings could be shredded or removed in a short time. Dung beetle communities are strongly influenced by dung type and they change according to availability. As such changes in community organization of dung beetles include changes in species richness, species composition, abundance and guild structure. In a total of just over 300 invididuals representing 21 species, the tunneler guild formed 15 species, the most abundant of the fonilds and the roller was least abundant. The most abundant species were *Drepanocerus setosus,* followed by *Onthophagus bronzeus* and *Liatongus indicus* (Sabu-Thomas, 2006). The elephant dung pats have high species richness of tunnelers, while a distinct pattern of dung resources partitioning is seen in the tunneler guild involving species of *Onthophagus* and *Copris signatis.* High unevenness in taxonomic structure of the guar dung beetle assemblage arises from increased presence of *Cacobuis* and *Onthophagus* species. ``The presence of 24 species of *Onthophagus* and three species of *Cacobius* in gaur dung as compared to the presence of 7 *Onthophagus* species and the absence of *Caccobius* is elephant dung is a noteworthy feature. In short, 65% of the species are attracted to gaur dung from the genera *Onthophagus* and 73% from the genera *Onthophagus* and *Caccobius.* (Vinod - Sabu, 2007).

Fig. 3.5. The rubber beetles, *Luprops tristis* (Courtesy Sabu Thomas)

Litter insect communities

The natural forest litter ecosystem supports a variety of macro- and microarthropods involving different trophic berth. Both species richness and abundance are characteristic, besides the degree of intraspecific diversity of fungal feeding thrips are unique aspects. As the litter layer become deeper it undergoes an orderly and sequential breakdown with faunal assemblages succeeding one another. Larger species of collembola such as *Tomoceros, Entomobrya, Orchesella* abound in surface litter, while the smaller *Onychiurus, Tullbergia* occur in deeper layers. The patterns of seasonal variation of collembola are diverse for diverse species. Seasonal fluctuation of some dominant species of *Cyphoderes, Brachystomella, Lobella, Lepidocyrtus* and *Folsomoides* is very typical. Several species of mycophagous thrips inhabit fungus-infested surface litter in natural forests. Species of such genera as *Hoplandrothrips, Hoplothrips, Ecacanthothrips, Elaphrothrips, Bactrothrips, Nesothrips,* are very typical of habitats such as drying, dead, fungus-infested leaves and twigs. Many of these species are associated with particular typs of fungus intestation, such as those of *Aspergillus, Pestalozzia, Lasiodiplodia, Fusarium* etc. Behaviourally some species tend to aggregate. *Preisneriana kabandha* laying an many as 10-110 eggs. *Ecacanthothrips* 25-40 eggs/female, litter inferting *Bactrothrips idolomorphus* 27-36 eggs with ovipary ovovivipary and vivipary.

Others like earwigs, psocids, cockroaches, termites, beetles occur in the soil. The larvae of carabids and staphylinids, coccinellids, click beetles are common, besides the diversity of termites enabling rapid turn over of organic matter in such ecosystems. Foraging on leaf litter, bark, dead wood are *Microcerotermes* and *Nasutitermes*. It is suggested that 50% of litter in forests in consumed by termites. Besides insects, oribatid mites are predominant inhabitat ants of forest litter with such common genera as *Galumna* and *Scheloribates*.

One cannot overlook the huge aggregation of the rubber plantation litter beetle, a tenebrionid from Kerala. Predormancy and post dormancy periods of this beetle *Luprops tristis* occur in rubber, while with the onset of summer rains, huge aggregations of adults invade residential buildings and enter into a state of dormancy for 9 months. Beetle aggregations are usually in the range of 0.5 million to 4.5 million individuals for residential building (Sabu *et al.,* 2008). Their massive seasonal invasions make sum the most dreaded beetle to farming communities surrounding rubber plantations.

Gall makers and leaf miners - Cecidogenous insects

Galls are abnormal growth produced by plants in response to the feeding and oviposition stimulus of insects, resulting in excessive growth and all multiplication. The specific habit of initiating a gall and exploiting if for nutritional and defense purposes, besides developing a progeny withinin, is a major feature, presenting a highly evolved form of phytophagy. Aphids, thrips aleyrodids, coccids, psyllids are among the well known gallers. The modes of gall induction differ significantly among the different group of gall formers.

Gall insects are of interest and are unique in that they can alter the plant tissues to form such growths. They are so closely involved with their respective hosts that they respond favourably to the action of the concerned insects, forming a nutritive zone which acts as a `sink' where all the essential metabolites required for insect development are present. The intimacy of association between the gall insect and the hort plant is a special biotic relationship. Involvement of two or more species of the same or different groups of insects, is a common feature and factors affecting gall size include the diversity of galls, number of larvae, and position on infested leaf. The size, structure and mutualism are under the control of the gall forming organisms and perfect coordination exists between the gall and the gall maker.

Gall forming thrips present a complex social behaviour and morphological variations associated with wing polymorphism. Several gall thrips abound in *Acacia*, some are opportunistic gall inhabitats, others are phyllode nest formers. Acacia thrips exhibit *pleometrosis*, the colony founding being by multiple adults, as in other social insects. The defensive nature of the gall habitat and the presence of `soldier' caste are of added significance in considering the social behaviour of thrips (Crespi and Mound (1997), not to mention of the multiple modes of gall thrips life histories and social systems. The galls range from flattened, hollow cylinders to spherical or kidney-shaped and hundreds of larvae are closely packed. Further details are provided in Chapter VI.

Fungal association inside plant galls

The fungus-insect-angiosperm association appears to have attained a unique dimension in galls of species of cecidomyiids. About 16% of the entire group of gall midges are exclusively mycophagous. Most species of *Mycodiplosis* feed on the mycelia of several species of mildew, moulds and rusts. The gall midge *Camptodiplosis* lives exclusively on the thalli of larger fungi like *Auricularia* and the larvae of *Brachyneuria* sp. cause small, irregular galls on the agaricine fungus *Peniophorae ovalis* producing small blister galls. Adults of *Lasioptera* complex and *Asphondylia* generally transfer the hyphae when the galls are initiated. This association is of interest in view of the trophic relationships of several other gall insects which induce galls primarily to draw nutriton from their respective angiosperm host tissues, the fungal mycelium establishing a luxurient growth within the gall, around the larva. In several cases as in thrips galls, the fungi invade the degenerating galls after the desertion of the gall insect.

Ectosymbiosis

One of the finest examples relate to leaf cutting ants which cultivate fungi and the infra buccal pouches of the ants is used as a storage organ. These leaf cutting ants use a variety of plant materials as fungal substrate. Species of the genus *Atta* and other fungus feeding ants manure this fungal garden with faecal material, which aid in directly supplying the needed aminoacids and other nitrogenous substances.

The ambrosia beetles including the scolylids and platypodids bore deeply into the wood of fallen trees and logs, cultivate ambrosia fungi in their galleries to feed the developing young. The beetles have special ectodermal pouches, called *mycangia* or *mycetangia* and the average number of such pockets varies from 68 in *Xyleborus affinis* to 160 in species of *Monanthrum*. Mycangia are pit-like and may be elytral, prosternal, prothoracic, pleural or oral (Batra, 1963). The number and distribution of mycangia, are of taxonomic value and tend to vary with sexes. Ambrosia fungi are extremely species specific in relation to insects and flourish inside wood tunnels throughout the active life of beetles, the ambrosia consisting of hyphae form a compact lining to the funnels. The fungi belong to such genera as *Lasiodiplodia, Monilia, Cephalosporium, Ceratocystis* etc.

Several species of mycophagous tubuliferan thrips have developed the tendency to aggregate, the nature of aggregation and population size mostly depending on the available fungal found. Ascomycetes, Coelomycetes and Hyphomycetes occur in dead and decaying wood and several species of thrips utilize the fungal food for feeding including several species of *Elaphrothrips, Hoplothrips, Bactrothrips,* to mention a few. Adaptive diversity in term of reproductive types are key typical of these thrips with oviparity, oviviviparity and viviparity being determined by factor such as environment, fungal food availability, aggregation and oviposition behaviour. The build up of fungal inoculum at the sites also enable efficient mechanical transport of these spores by their association with thrips species (Ananthakrishan, 1989).

Extrafloral nectaries occurring on stem, bracts, stipules, flower buds, peduncle or external floral surface harbour diverse species.

Being highly nutritious they serve as phagostimulants to insects. Extrafloral nectaries are sugar producing glands occurring more frequently in Leguminoseae, Orchidaceae, Euphorbiaceae. The ecological importance of nectar-feeding ant colonies varies from plant to plant. Diverse natural enemies feed on them, such as adult and larval lady bird beetles, parasitic and predaceous wasps, tachinids and syrphids, lacewings, carabids and mantispids. A single species *Vicia faba* has recorded 60 species of parasitic Ichneumon wasps imbibing nectar (Whitman, 1994). Extrafloral nectaries make an indirect positive contribution to pollination, by keeping away harmful visitors.

Domatia

A wide variety of ants make use of domatia derived from various plant parts such as hollow roots, hollow stem, hollow inflated stem sheaths, leaf, stipule or petiole pouches, hollow, stipular spines and swollen interlocking spines (Whitman, 1994). Such living plant tissues are regularly occupied by ants, called *myrmecophytes*, with complex adaptations to house and feed specialized ants. Ant inhabited acacias harbour 750 species worldwide and swollen thorn acacias have extra floral nectaries at the petiole. Plant structures regularly inhabitated by ants are called *domatia,* a good example being *Myrmecodia tuberosa,* an epiphytic shrub, with enlarged tubers containing large, empty cavities having a variety of ant species. Seed dispersal by ants or *Myrmechory* is common in Australia and Africa, the number myrmichorous species estimated to the more than a thousand species. Interestingly, most ants consume seeds and are called `harvester ants' with such genera as *Monomormim, Pheidole, Pogonomyrmex* which tend to build granaries in which large numbers of seeds are stored.

Mutualistic Interactions

With the rise of evolutionary ecology, the evolution of adaptation within species became intertwined with the evolution of such interactions. Different kinds of ecological specialization are identifiable, such as the wide tolerance of insects to diverse ecological conditions, genetic polymorphism within population of individuals adapted to particular niches, ecotypic variation among populations and specialization for narrow niches, are known. Selection for specialization tends to be limited by the kinds of resources they exploit. Mutualism yields reciprocal benefits at individual/population levels and may be obligate or facultative, symbiotic or non-symbiotic and diffuse or tight. Homoptera infesting a wide range of plants exude honeydew which attract a number of ants and whiteflies associated honeydew and wax, the white flies benefitting from ant tending, because the presence of ants deter the onset of predation and present build-up of fungi collected in the honey dew. Of particular interest are the lycaenid butterflies which secrete honey-like fluid on their surface and ants that protect the caterpillars from other predatory ants, wasps etc. The diversity of such mutualism in insects is common and the high frequency of mutualism in communities depend on the richness of antagonistic interactions. Because mutualism tends to create new resources, they can become foci for the development of new interactions, contributing to the complex network of interactions and organization of communities (Thompson, 1982). Multitrophic level interactions have come to be recognized, providing a holistic picture to insect-plant interactions.

Interactions between individuals of different species are among the most important processes influencing the patterns of adaptation and variation of species (Thompson, 1994) and of community organization. With the rise of evolutionary ecology, the evolution of adaptation within species became intertwined with the evolution of such interactions. Different types of ecological specialization are identifiable, such as wide tolerance of individuals to ecological conditions, genetic polymorphism within populations of individuals adapted to particular niches, ecotypic variation among populations and specialisation for any narrow niches, are known. Community specialisation limited by the kinds of resources they exploit and specialisation of hosts to one visitor species, by the tendency of mutualistic resources to attract new visitor species. In short, interspecific interactions are one of the most important processes influencing the patterns of adaptation and variation in species (Thompson, 1994) and of community organization and stability. The nature of such interspecific interactions, structure of food webs are involved in studies of mutualism involving multispecies multitrophic interactions. For example, the spatial and temporal patterns of community structure, such as seen in ant communities enable a better appreciation of how species interact within communities as networks each species being connected to one or more species (Rico-Grey and Oliveira, 2007).

As regards the exploiters of mutualistic interactions, mutualism yields reciprocal benefits at the individual population levels and may be obligate or facultative, symbiotic or non-symbiotic and diffuse to tight. Hemiptera including aphids, coccids, membracids etc. exude honey dew which attract a number of ants, and whiteflies are frequently associated with honey dew and wax, the white flies benefiting from ant-tending, because the presence of ants deters the onset of predation from other host plants and prevent build-up of fungi in the honey dew. Of particular in that are the by lycaenid butterflies which secrete a honey-like fluid on their surface and ants protect the caterpillars from other predators. The diversity of such mutualism in insects is common and the high frequency of mutualism in communities depends on the richness of antagonistic interactions. Because mutualism tends to create new resources, they can become the foci for new interactions contributing to the complex network of interactions and organization of communities (Thompson, 1982).

Ant-plant interactions are suitable systems for an understanding of the ecology of terrestrial insect communities, more especially the processes operating at the community level. The usage *ant-biomass paradox*' suggests that foliage dwelling ants behave mostly as primary consumers and depend mostly on plant and insect derived nutrition (Tobin, 1991, 1994). More than 86% of the arthropod biomass and upto 94% of the arthropods live in the canopy of Tropical forests, by their ability to process liquid food in their defensive system. The high incidence of exudates on foliage play a key role in shaping food web structure in Tropical forests. The continued availability of liquid food on foliage plays a key role in mediating foraging ecology and community organization of canopy ants (Rico-Grey and Oliveira, 2007).

Functional diversity of ants

An ecological community is a collection of species living in a given environment and are grouped into those related to form, function and diversity. Given their diversity and biomass, they play a large role in the functioning, of ecosystems. Ants affect the structure of their environments as ecological engineers i.e. organisms that rearrange the environment in ways that can affect other organisms. Ant functional groups are known, such as climate specialists such as hot, cold and tropical, opportunists, specialized predators and the like. Interaction of ants relate to parasitism, commensation, nutrition. Leaf cutting and homopteran tending ants acquire their through symbiotic fungi and homopterans. Evolutionary praecursors for mutualism in which plants receive protection from herbivory, in exchange for supplying cavities favouring ant occupation hollow branches, stem and thorns, pouch-like domatia on leaves and petioles, providing food as well as shelter; encouraging frequent visitation of ants. Some ants occupy and cultivate ant gardens ie clusters of epiphytes planted in the branches of trees. Ants like *Solenopsis, Crematogaster, Solenopsis* obtain nutrition from extrafloral nectaries. Ant-homopteran associations are very well known involving interactions with aphids, cercopids, cicadellids, coccids, membracids, fulgorids, Ants feeding on honey dew, a by-product of homopteran phloem feeding, consisting of carbodydrate and amino acids added by the homopteran. In return the ants protect the homopterans from predators/parasitoids.

Fig. 3.6. Left: Flower infesting thrips; Right: Leaf miner attack.

Fig. 3.7. A.B. Coccids and ants; C. Aphids and ant feeder, D. Scales and ants.

Fig. 3.8. Membracids and ants
(Courtesy Dr. Prabhakar, ZSI, Chennai)

Ants as mutualists

Ants comprise an important group of mutualists for reasons more than one. (a) they inhabit virtually every environment, within ecosystems with large scale population (b) the social structure of ant colonies enhance this effectivenee as mutualists (c) indirect interaction such as ant nest building activity turns the soil, thereby benefitting particular plant species. Such mutualistic interactions between ants and insects like Homopterans are known as *trophobiosis*. Ant tending is also associated with high rates of honey dew production, which may attract predation or facilitate fungal infection. There is a positive correction between honey dew production and survival, between life span and life time fecundity for untended aphids. In the presence of ants, high quality aphids are tended more intensively by ants, which may improve their survival.

As for the benefits of ant tending, honey dew is an important source of carbohydrates for ants. As for aphids tended by ants, they are able to feed rapidly and hence reproduce at a greater rate. Ants may move aphids to new areas for colonization. Ant adaptation is also behavioural, since they respond to aphid alarm pheromones with antipredatory behaviour, transport aphids to new breeding sites and new host plants. Aphids may be more able to aggregate when tended by ants. In short, ant-Homoptera relationships are mutualistic in view of the tending ants harvesting the energy rich fluids.

Mutualistic association between ants and Hemiptera are diverse, complex multispecies systems. Seasonal patterns of host plant use by Hemiptera like aphids particularly the selection of feeding sites are common. For instance, aphid colonies on floral tips of *Nerium* attracted three to four times as many ants. ``Host plants may modify the potential for ant-tended insects to attract ants and thus influence the formation and outcome of interactions between ants and exudate producing insects (Price *et al.*, 1980; Strong, Lawton and Southwood, 1989; Cushman (1991) and Rico-grey and Oliveira, 2007). The presence of ants may increase the fitness of the hemipteran by influencing the location of feeding. Interspecific interactions in short, are one of the most important processes influencing the pattern of adaptation and variation in species (Thompson, 1982; 1994).

The remarkable ecological success of canopy ants are attributable to their ability to overcome some of the constraints of this particular environment including a limited number of nesting sites, decreased amount and diversity of food resources, low humidity and strong limits (Rico-gray and Oliveira, 2007). The widespread occurrence of liquid food resources such extrafloral nectar, honey dew from sap-feeding hemiptera are far more abundant. The predominance of ants is explained as due to the occurrence of liquid food on foliage and the high abundance of such resources on foliage plays an important role in shaping the web-structure. Such ants also provide substantial nitrogen from live prey.

It would be worthwhile examining a few instances of mutualistic interactions such as whiteflies and ants; mealy bugs and ants; ants and membracids and ants and aphids.

White flies and ants

Considerable mutual relationship between aleyrodids and ants are evident, the former dependent on the plant sap with nutritive substances and in turn, the ants tending them are attracted by the sugary fluid released by the aleyrodids. Interactions between the white fly *Rachispora ixorae* infesting *Ixora pavetta* show that these white flies are dependent on the plant sap, digested by the hemipteran which emit a sugar-rich honey dew through their vasiform orifice as drops and this attracts different ant species which use them as stable source of nutrients. Both biotic and abiotic factors influence the white flies and ants enabling a mutualistic relationship. The whiteflies receive protection from parasitisation and predation, whereas they in turn provide the ants a constant supply of food enabling a trophobiosis relationship. They are also able to minimize the deleterious effects of fungi by removing excess of honey dew.

Mealy bugs and ants

Ant tending increased densities of the mealy bug *Pseudococcus* and lowered densities of parasitioids. Densities of the coccinellid predator *Cryptolaemus* were higher on ant-tended vines where there was more mealy bugs. Attending ants belong to the species *Crematogaster, Dolichodera.* The black ant *Camponotus* besides foraging on the nectar of various plants, also tended several homopterans in orchards. *Camponotus compressus* was seen to tend *Pulvinaria* in large numbers. While the ants forage for honey dew, the only benefit scale insects derive in the colony hygeine through constant removal of honey dew.

Ant and membracids

Several species of membracids are tended by ants and such associations are common on thorny Acacias. Ant-membracid association positively affect plant fitness if the benefit from ant derived protections from herbivory, outweights the losses incurred by membracid feeding. Tree hoppers benefit from ant attendance through protection from predators. Membracids in small aggregation benefit more than tree hoppers in large aggregations indicating a density dependent benefit. Ant effect on tree hopper fecundity also varied with time and shift of natural enemies. Ants receive energy rich food and in turn increase tree hopper survival and reduce damage to the host plant. Instances are known membracids benefit more when tended by ants like *Myrmica.*

Ant-aphid association

Ants can have simultaneously both positive and negative effects on aphids. The relative effects of predation and production by ants on aphid abundance show that ants have a strong negative effect on aphid numbers. When attacked by predators, aphids secrete the alarm pheromone which cause them to disperse. Ant-associated aphid species disperse less effectively than non-myrmecophilous species. Ant-tending is associated with high rates of honey dew production.

Ant-plant interactions have played a major role in our understanding of mutualism, offering an excellent opportunity to study impact of ecological factors on the evolution of such interactions, enabling better opportunities to be used as model systems in ecological and evolutionary interactions.

Chapter - IV

Insect communities in Forests and Agroecosystems

Natural forests, both pristine and degraded, have in recent years been subjected to the large scale introduction of many fast growing exotic species resulting in increased pest incidence. To quote Berryman (1986) ``the World-wide distribution of forest trees is being continuously changed as exotic species are used more and more in plantation forestry.... we should expect trouble from insects in these exotic plantations". Native or indigenous species are adapted to the local environment and evolved considerable resistance to local pest and disease. Being ubiquitous components of terrestrial ecosystems, natural forests and herbivorous insects contribute to the cycling of nutrients and energy in the ecosystem. An undisturbed ecosystem tends to have communities that are very diverse in species, but also stable, in the sense that loss of species is balanced by periodic replacements. Such communities have well established links between coexisting species as evident by the presence of food chain or food webs.

Plantation forestry is the most common forestry practice, replacing species-rich natural ecosystems by the species-poor monoculture by them of trees for purposes of commercialisation. Ecological inputs are important in management and knowledge of tree biology and architecture, patch dynamics, ecophysiology of developing forest communities, reproductive biology and nutrient cycling are aspects to be inegrated into management practices (Ramakrishnan, 2002). With the increasing disappearance of Tropical forests, the implications of biodiversity are considerable. Many ecologists have considered Tropical forests as constituting a comparatively stable and highly coevolved sets of niche-differentiated species persisting in a kind of equilibrium in the community (Connell, 1978). The coexistence of a large number of tree species in Tropical forests is also perhaps due to predators of seeds by host-species herbivores, producing a uniform dispersion of mature trees (Hubbel, 1980). However, one has also to keep in mind that ecosystem, communities, species and populations lack clear boundaries in space and time.

Outbreaks of pests do occur in the ecosystem associated with high a high degree of host tree density such as that of the defoliation *Hyblaea peura* in teak, *Hoplocerambyx spinicornis* on *Shorea robusta,* and the bag worm, *Pteroma plagiophleps*. Insect outbreaks such as that of stemborers and sap feeders are common due to adverse conditions and a notable example is the scolylid bark beetle, *Dendroctonus frustatus* (Nair, 2003). An increase in the number of borers attacking a host tree defence effectively cause increase in populations or outbreaks. The causes of such outbreaks include dramatic changes in the physical environment, changes in the genetic component of pest populations, qualitative changes in host plants and life history strategies of pest species (Berryman, 1987). Exponential growth of populations, competition for resources, complex web of interactions with other biological populations could lead to unpredictable population explosion. As such, populations could be stable, cyclical or eruptive. Many polyphagous insects in forests have stable populations, while cyclical populations involve regular cycles of abundance, with eruptive species occurring sporadically, like *Hoplocerambyx spinicornis* (Nair, 1993).

In plantations several factors promote increase in pest populations with monoculture plantations being more prone to insect attacks, so that pest problems are greater in plantation forests. Factors favouring pest build-up in the exotics include fewer competitors for the resource and absence of specialized natural enemies in the case of introduced pests. Insect species attacting exotic tree species appear to increase overtime such as that on *Leucana leucocephala. Pteroma plagiophleps,* the bag worm has over time developed into a major pest of exotic plantation of *Paraserianthis falcatoria* and has shown the capability of developing host races and can over time become serious pests. Large increase in the number of individuals of a species creates pest situations, notably increase due to host abundance, host quality improvement and formable conditions. Uncontrolled increase of populations resulting in outbreaks can occur, such as increase in the number of borers attacking a host tree, causing a further increase of population, leading to a chain reaction resulting in uncontrolled increase of outbreaks such on that of the teak defoliator *Hyblaea peura* in new teak plantations.

Both exogenous and endogenous factors cause population fluctuation. Three basic types of population behaviour include relatively stable population change, exponential growth of populations and cooperation among individuals of the same species, competition among individuals for resources linking population to environment such as resources and complex web of interactions with other biological populations (Berryman, 1995). Different causes tend to lead to the same type of spatial growth or behaviour, such as relatively stable population or eruptive population. Most polyphagous insects belong to relatively stable populations.

In plantation forests however, abundance of food is ensured creating a very large extent of area facilitating build-up of pest population. Accumulation of wood residue during site clearance tend to favour build-up of borer insects (Nair, 2003). Trees in poor sites may act as epicentres for the build-up of some pests. While natural populations show high genetic variability, plantations are less diverse and are susceptible to pests. Sometimes species such as *Tamarindus indica* are infested by *Pteroma plagiophleps* which has over time developed into a serious pest. Similarly the spindle bug of arecanut *Caravalohoia arecae*, an important pest occurs in large colonies. In short, the factors favouring outbreaks in plantations, as outlined by Nair (2003) are :-

(a) the basic determinant of population size is abundant food

(b) accumulation of wood residue

(c) stressed trees as epicentres for build-up of pests

(d) Closeness of trees in plantations

(e) Survival of natural enemies facilitated by absence of other plant species.

(f) changing nutrient composition of trees promoting growth and development of some pests.

(g) plantations are generally less diverse with genotypes more susceptible to some pests.

As has been indicated earlier, factors favouring pest build-up in exotics and fewer competitors for the resource and absence of specialised natural enemies. Further, the likelihood of establishment of major pests from their native habitats are remote. Tropical forests are amongst the most productive ecosystems and species can afford to specialize on a chain of resources enabling easy competition for their chosen food. The discovery of thousands of beetles and other insects in forest canopies is astounding and it is not an exaggeration to indicate that nature reaches its fullest expension in Tropical forests, both by the number and complexity of interactions. Intricate predator - prey

Fig. 4.1. Pentatomid aggregation in bark (Top); Teak grasshopper nymphs (*Aularches scabiosus*) destroying teak leaf (Courtesy: V.V. Ramamurthi)

webs are common, leading to interspecific interactions. The vertical dimension in tropical forests is able to accomodate high levels of diversity and to give one example, from the herbaceous ground layer having large contingent of ant species to the far richer canopy diversity is considerable. Gregarious species feed more and rest in groups. Aggregation could help thermoregulation and defence. Gregarious caterpillars tend to overcome host plant defence through mass attack or may induce changes in the host because of feeding in a limited area. Some species of caterpillars are genetically programmed to aggregate, showing well defined patterns of integrated behaviour. There is a wide range of responses among the hundreds of species in natural forests to changes of forest quality leading even to the disapearance of the more sensitive species. It is not only the loss of habitat quality, but also of the reduced availability of resources in such degraded forests, that are responsible for the reduced population size. As such the four dynamic processes that limit population size are *natality, mortality, emigration* and *immigration*. Further, the loss of microhabitats results in the loss of habitat quality, and to quote Hanski (2006). `Reducing habitat area inevitably cuts off some parts of the variation in the environmental condition that used to buffer the population dynamics against variation in environmental conditions".

Fig. 4.2. Psyllids on *Leucana leucocephala* (Courtesy, Prasanth Jacob)

To quote McNeely (1996) ``It appears that the best way to maintain biodiversity in forest ecosystems is through a combination of strictly protected areas, multiple use of areas managed by local people and natural forests extensively managed for suitable production of commodities such as logs and forest plantations intensely managed for the wood products needed by society". Basically there are two situations where insect outbreaks occur, first when they attack weakened trees, building up large populations capable of successfully attacking healthy trees. The second situation is a density related one in which vast amounts of food and favoured conditions are present. Outbreaks of insects though tending to cause heavy mortality are different, in the sense that insect feeding is more selective in terms of tree species and takes place over a long periods of time. Again there are insects like defoliators, sucking insects, bark beetles which have catastrophic effects producing different ecological effects. Fresh dead wood with intact bark are attacked by cerambycids or longicorn beetles, while smaller beetles such as anthribids, curculionids and lyctids attack wood devoid of bark. Ambrosia beetles with fungal association help in the degradation of cellulose and lignin and fungal decay promotes infestation by termites whose density, in Tropical forests range from 390-4450 million individuals/m^2 (Sen Sarma, 1998). In short, the need for a better understanding of insect communities, is a pre-requisite for an overall generalisation regarding the structure and function of diverse species involved in forest ecological studies. In this connection it may not be out of place to cite a few well known examples.

Fig. 4.3. (Top) *Eutectona macheralis* Larva; (Bottom) *Hyblaea puera* Larva (Courtesy, Prasanth Jacob).

(a) *Gmelina arborea* infested by leaf feeding *Eupterote undulata* and the stem boring *Euwallacea formicata.*

(b) *Leucana leucocephala,* The mimosaceous introduced plant used as fodder, green manure and firewood.

Heteropsylla cubana builds up in large colonies feeding on the terminal shoot which dries up and stunting growth.

(c) *Swietinia macrophylla*, a valuable timber tree Mahogany. The shoot borer *Hypsipyla gradiella* a pyralid, the larvae boring into the growing shoot, destroying the terminal bud, causing retardation of growth and forking of stem.

(d) *Tectona grandis* or Teak.

Hyblaea peura, the leaf defoliator, *Eutectona machaeralis,* the leaf skeletonizer; *Xyleutes ceramicus,* the stemborer and *Asphondylia tectonae*, the cecidomyid are noteworthly pests, causing significant damage to teak plantations.

Fig. 4.4. (Right) First instar having of Bihar caterpillar; Whitefly nymphs on cotton leaves (Left).

Fig 4.5 Thrips colonies.

Interestingly, in the living teak tree, 187 species have been on record including leaf feeders, sap feeders, fruit and inflorescence feeders, stem and rot feeders, with 79 species of Lepidoptera and 40 species of Coleoptera (Mathur and Singh, 1960).

(e) *Paraserianthes falcataria,* a Mimoceae fast growing tree earlier known as *Albizzia falcataria.* 40 species of pests are known of which *Pteroma plagiophleps,* the bagworm, the moths *Semiothesia* and *Archips,* the membracid *Oxyrrachis tarandus* are well known. The Hepialid moth *Sahyadrassus malabaricus* feeds on saplings, boring into stem and is also a bark feeder.

(f) The seed feeding brachid *Bruchidius* sp.

Impacts of Changing habitats on Insect Communities

Natural ecosystems are endowed with high species and genetic diversity with heterogenous habitats harbouring species rich insect and plant communities. Increased resource exploitation and alteration of land use pattern have resulted in the fragmentation of habitats and ecosystems, reducing colonization and population sizes. Survival of species is invariably affected by fragmentation of ecosystems, of biotic communities, of species population and habitat range of a particular species. Different species have different ecological requirements making habitat deterioration vulnerable to species reduction or loss. The loss of bark and litter communities of thrips due to deforestation and logging in areas like, the Tenmalai forest of Kerala, harbouring a wide variety of trees such as *Messua, Dipterocarpus*, *Barringtonia*, *Kigelia*, to name a few, constituting specific microhabitats. The composition of mycophagous communities in such areas is diverse with species of genera such as *Ecacanthothrips, Polyphemothrips, Dinothrips, Elaphrothrips, Bactrothrips, Nesothrips,* to mention a few. Species composition of the once natural forests has dwindled with deforestation and gradual fragmentation of habitats. In short, the disturbed landscape tends to cause perturbations leading to a chain of events leading to the extinction of several species of insects. Patches of rare insect communities are more susceptible to disturbance leading to increasing loss of genetic variation. Large tracts of forests which may be fragmented, result in increased numbers of edges, having implications for being susceptible is increased distribution of weeds and other plant species and associated insects. The overall insect diversity in the Tropics is substantially affected by land use management, both in agriculture and forestry. Ants, lepidopteran larvae, beetles such as carabids are often used as indicators of land use changes. Needless to emphasize that fragmentation of habitats with smaller, isolated patches present a severe threat to poorly dispersing species. Further, their very existence of a particular species depends on the presence of some other species and the extinction of one species may lead to the extinction of one or more species resulting in a chain of extinctions. Brooks *et al.* (2003) estimated that Singapore lost 40% of its butterflies in response to 95% loss of in forest habitat in 183 years.

Forest insect studies in short, need to document patterns of diversity and community stucture. All species are embedded in complex webs or mutualistic and antagonistic interactions and no where are these webs more complex and diverse than in tropical forests (Janzen, 1983). Trophic interaction are the glue that hold together ecological communities (Lewis and Bassen, 2007). Small scale farming is an important cause of forest degradation notably of commercial logging. The author had occasion to witness such changes in certain areas of the Western ghats, wherein forest lands were utilized for tapioca and rubber plantations and the loss of several mycophagous thrips species associated earlier with natural forest sites. The need to understand how the structure of tropical communities is affected, what are the acceptable limits of tropical forest biodiversity and the groups of insects seriously affected, are aspects which will go a long way in preserving biodiversity in natural forest ecosystems.

Community dynamics in agroecosystems

Agroecosystems are dynamic, multifacted evolving complexes with continuous interactions between insect populations and their host plants. Crop intensification leads to the disturbance of the balance between host and insect communities. Specialist and generalist insects evolve with crop diversity and a knowledge of insect population dynamics, their tendency to aggregate in different crops and impact on plants have become essential aspects of economic entomology. In short, agroecosystems comprise the total complex of organisms in the crop area in relation to the diversities of the environment. The components of agroecosystems tend to fluctuate, with the kinds of crops, agronomic practices, changes in land-use pattern being important elements affecting the stability of agroecosystems. Land races have evolved or developed forming basic raw materials, so that crops and weeds have become interwoven, in the sense, that such weeds become reservoirs of several pest species which migrate to the crops, resulting in wild species becoming gradually transformed into cultivated forms. In short, agroecosystems are ecological and socio-economic systems comprising domesticated plants and animals and people who sustain them for purposes of producing food, fibre and other agricultural products. Agricultural technologies that have high potential for sustainability are intercropping, rotation, agroforestory or annual herbaceous crops interspersed with perennial trees or shrubs. Productivity, stability, sustainability and equitability are the four pillars of agroecosystem development, with biodiversity and ecosystem complexity contributing in diverse ways to ecosystem functioning.

Intensive and extensive cultivation of high yielding varieties, along with with the application of increasing numbers of agrochemicals have resulted in the aggravation of pest problems with many hitherto known species assuming the role of serious pests, which have also developed resistance to pesticides. Ecological implications of large populations of insect pest species need to be assessed, particularly intensity of attack and resultant impact on crops. Spatially explicit patterns of differential damage by insects usually have a demographic, distributional and potentially evolving consequences.

The influencing factors of pest infestation are so many that an estimate of the exact situation is hardly possible, though it is a function of insect population density. The critical population level of the pest depends on the stage of the plants, locality, type of pest and association with other organisms. Assessment of economic injury levels is an important aspect and sociological, ethical, environmental and economic aspects become necessary to understand the nature of economic damage. Consideration of the economic threshold of a pest population, namely the larva at which a pest population reaches, the economic injury level, becomes important. The general level of the abundance of a species, the kind of damage caused and the damage potential are important aspects, besides their role as occasional, potential or migrating pests are equally important. However for each agroecosystem their exists a few key pests when the crops are considered alone. For example, taking the insect pests of sorghum, nearly 150 species have been reported as pests involving shoot borers, stem borers, armyworm, shoot bug, aphids, grasshoppers and locusts, mirid bug and panicle feeding caterpillars (Sharma and Nwanze, 1997). Taking only their egg-laying capacities, it is known, that wireworms such as *Gonocephalum, Eleodes, Conoderus,* lay as many as 50-300 eggs; stem borers *Chilo, Sesamia* and *Diatraea* lay as may as 500 eggs; armyworm *Mythimna, Spodoptera* lay 500-900 eggs; *Schistocerca gregaria* egg in masses of 20-200; each female aphid

Schizaphis graminum gives birth to 60-100 nymphs; tea bugs *Calocoris angutsatus* lays 150-200 eggs; panicle feeder *Nezara viridula,* 300-500 eggs; *Helicoverpa armigera,* laying 350-3000 eggs. The need for more information on population fluctuations, carry over from one season to another and the role of biotic and environmental factors on insect development and abundance cannot be over emphasised. Non-residents in an ecosystem are the migratory pests-apprearing all of a sudden on a crop causing heavy damage, their association with the agroecosystem being a transitary one. High potential rate increase, strong dispersal and host finding are the r-pests like locusts, aphids etc. while low rates of potential increase, great interspecific composition ability, more specialised food habits are the k-pests which suddenly become pests of economic importance. Most of the crop pests come in between the two categories are the intermediate pests.

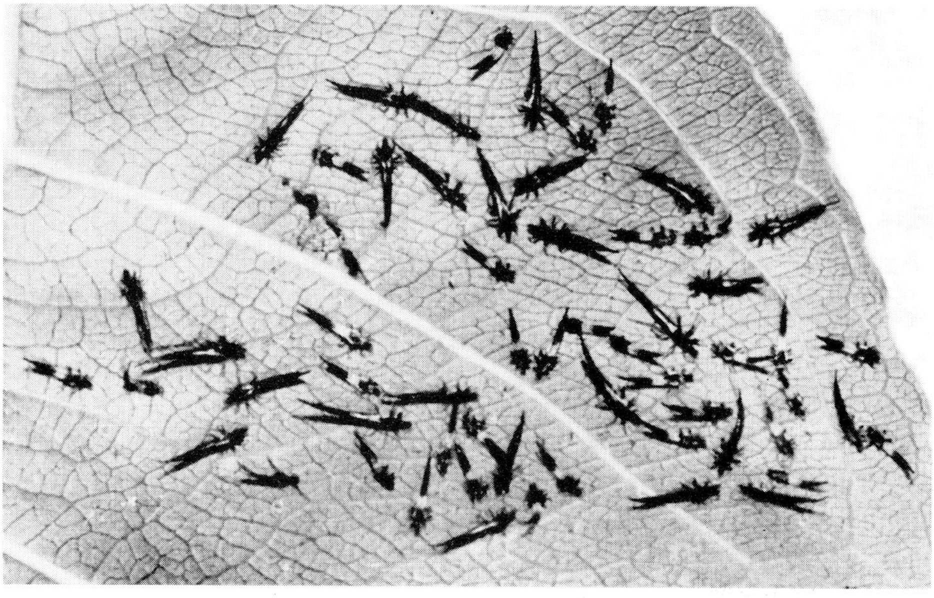

Fig. 4.6. A colony of *Cercothrips nigrodentatus* on Ficus leaf.

Agroecosystems as such are dynamic, multifacted evolving complex with continuous interactions between. The population of insects and their host plants, comprising the total complex of organisms in the crops area is relates to the diversites of the overall environment. The components of agroecosystems tend to fluctuate, the kinds of crops, agronomic practices, changes in land use patterns are important elements affecting the stability of agroecosystems. Locusts and army worms are well known migrating pests, while there are the potential pests which tend to cause problems in certain situations. For example, the larvae of *Spilosoma obliqua, Amsacta albistriga, Spodoptera litura* are gregarious in the early larval instars, not to mention of coccids, aphids, plant hoppers such as *Amrasca biguttula* and *Empoasca flavescens.* Insects like aphids, aleyrodids, coccids, psyllids, thrips by their gregarious habits tend to cause adequate damage. Besides, several species of thrips, aphids, psyllids, coccids also cause galls in which hundreds of individuals occur. The mustard aphid *Lipaphis erysimi* alone has been known to cause yield losses varying ton 35-73% in *Brassica.* In pulse crops the pod borer *Heliothis armigera* causes yield losses from 4.2 to 39.7%, while it may be as high as 50% in pigeon pea and around 30-35% in black gram (Dhaliwal & Arora, 1996).

Fig. 4.7. Aphid colonies on *Phyllanthus emblica* (after Prasanth Jacob)

Huge populations of *Mallothrips indicus* on *Syzygium jambolanum* literally overflow tree trunks, branches and leaves while a single flower of *Ruellia tuberosa* can harbour 820 to 1078 individuals of thrips, while bamboo leaf clusters are known to harbour 14 species with around 800 individuals (Anthakrishnan, 1984). Thrips community studies within inflorences of Compositae reveal seasonal incidence with species packing as evident in the weeds *Tridax, Wedelia* sp., invloving species of *Haplothrips tardus. Haplothrips gowdeyi, Frankliniella sulphurea* and *Microcephalothrips abdominalis.* By far the maximum number of individuals within a host species is in *Calycopteris floribundus* harbouring 10000 individuals within galls. In so far as major pests of crops are concerned, such species as *Amsacta albistriga,* the red hairy caterpillar, the larvae in literally thousands feed gregariously and march from field to field; *Opisina arenosella* larvae feed gregariously within galleries of silk and grass on the green matter of the lower surface of coconut leaflets. Similarly the Bihar hairy caterpillar *Spilarchia (Spilosoma) obliqua* lays over 2000 eggs and larvae feed in groups. *Cercothrips nigrodentatus* literally crowd the leaves of *Ficus* sp., while *Retithrips syriacus* aggregate in large groups on the lower side of castor leaves. In short, individual crop fields should be considered as islands in a matrix composed of a mosaic of habitats and insects invade and reinvade the fields from the surrounding matrix (Prince & Waldbauer, 1982). Changes in plant architecture and landscape diversity as a result of agricultural practices tend to strongly influence the presence and distribution of insect species. Further, in natural ecosystems, changes in one spease as a result of selection pressure, can produce subsequent changes in other species which interact with it. Multiple cropping systems have an important advantage. When two types of crops are grown in a field, the pests of one type are sometimes kept down by predators that inhabit the other. Such beneficial interactions can be useful, when the crops are rotated regularly. Pests, inclusive of weeds insect, pathogens cannot adapt themselves to a single set of environmental conditions and therefore do not increase quickly.

Fig. 4.8. *Mallothrips indica* larvae on *Eugenia jambolanum*

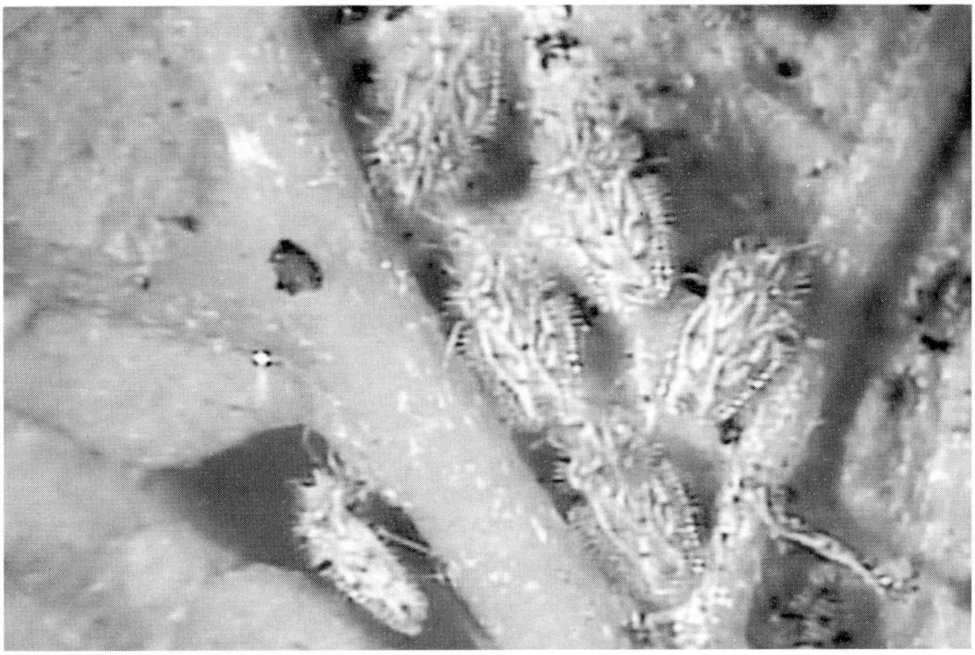

Fig. 4.9. Brinjal Tingid bug *Grentius hystericellus*

In short, agroecosystems do harbour a great deal of biodiversity, valuable for human welfare. Expansion of monoculture crops and decreasing habitat diversity have led to more insect outbreaks. The large size and homogenecity of crop monocultures can directly affect the abundance and diversity of natural enemies. In natural ecosystems the assemblage of insects is the result natural selection, but agroecosystems consist of unnatural assemblages of species and the opportunistics species frequently constitute weed, insects, and pests of disesase. Needless to emphasize that reintroduction of biodiversity with large scale monoculture is by establishing vegetationally diverse field margins that may serve as biological corridors allowing the movement and distribution of useful insect biodiversity (Altieri and Nichols, 2002).

CHAPTER - V

Biological and Ecological Inputs in Relation to Insect Sociality

Social life has been highly successful in the evolution of insects, being ecologically pre-eminent, with ants, and termites occupying the central stage in terrestrial environments. To quote Wilson (2005), ``Social insects are the ecological centre and solitary insects at the periphery''. Eusocial insects are the most advanced, involving care of young two or more overlapping generations in the same nest and existence of a reproductive and non-reproductive worker caste, so well evident in honey bees, mound building termites and army ants. Sub-social behaviour in evident in some Hemiptera which protect young from predation, in scolytid beetles guarding their young and feeding them with fungi, besides several mycophagous Tubuliferan thrips with extreme ocdymerous males guarding the larvae. The myrmicine ants are very diverse, with various types of specialists from exclusive predators to seed harvesters and aphid pastoralists. Many others like Pseudomyrmicinae are tropical arboreal ants showing mutualistic symbiosis as in ant-plants or myrmechory. The only ants to use formic acid as a weapon, are the myrmicines, which act as predators or scavengers. The complexity of communication systems and social organization are met with in honey bees, *Apis mellifera* being the best known. The stingless bees *Melipona* and *Trigona* are equally well known for their social organization. Besides these, are the bumble bees, Bombycinae. Wasps are also eusocial including the paper wasp *Polistes,* besides hornets and vespulids which have simple brood care without castes, to well established eusociality and strongly differentiated queens. Termites have a complex caste system and wingless workers relying on chemical communication through recruitment pheromones, alarm pheromones and *Trophallaxis* or exchange of liquid food and transmission of chemicals for communication, and attaining a large colony size, building termite mounds. Termites are chiefly decomposers and soil movers of the environment. Mound building termites, *Nasutitermes* and *Macrotermes* are among the most populous, raising fungi to assist in cellulose digestion. Soldiers are major workers meant specially for defence. Age polytheism or changes in behaviour with age is common, with a foraging tendency to return to brood care of nurses are few, a young nurse worker being able to take to foraging very early (Wilson, 2005). In short, eusociality occurs in ants, bees, wasps and termites. Such eusocial insects exhibit overlap of generations, cooperative brood care and reproductive caste differentiation.

Fig. 5.1. *Ropalidia cyathiformis*. Locality: Indian Institute of Science Campus (Courtesy R. Gadagkar)

Another category refers to the *primitively eusocial* wasps, where the queen and workers are not differentiated, with some individuals succeeding in becoming reproductive queen, while others end up in sterile workers, as is typical of *Rhopalida marginata* which have generally small colonies with queen and workers not morphologically differentiated, caste differentiation is post-imaginal and queens suppress workers by physical aggression. Besides, *R. marginata* is another paper wasp *Rhopalida cyathiformis* which is also primitively eusocial, without a well differentiated queen, all looking alike, the number in a colony rarely exceeding a hundred. The wasps fight and the winner becomes the queen for a while, since it is driven out by others which become the next queen. Individuals which are not queen act as workers, do not reproduce, but build the nest. Any wasp can start a nest and bring up offering. The work of Gadagkar and colleagues on *Rhopalida* has contributed immensely towards our knowledge of the primitive eusocial wasps. Interestingly, in contrast to *R. marginata*, in the eusocial *R. cyathiformis,* the queens are the most dominant, active and interactive individuals, suppressing worker reproduction and regulating workers' foraging. Upon removal of a queen, a potential queen with levels of aggression even higher than that of a queen, becomes immediately apparent. Such a potential queen appears to take over initiation of worker reproduction and regulate worker foraging (Kardile and Gadagkar, 2003).

Fig. 5.2. *Ropalidia montana* (after R. Gadagkar)

Fig. 5.3. *Ropalidia marginata* (after R. Gadagkar)

Fig. 5.4. *Ropalidia variegata* (after Gadagkar)

Fig. 5.5. *Oecophylla smargadina* (after R. Gadagkar)

Equally fascinating is the ponerine ant *Dicamma ceylonense,* a queenless ant species with the females with club-like thoracic appendages called *gemmae*. In the absence of a winged queen, the mated workers or gamergates, one of which function as a queen and mutilates all workers who eclose after her, by chipping their gemmae, making them incapable of sexual mating (Agrahari and Gadagkar, 2003). In short, primitive eusocial species have small colonies, queen differentiation is post-imaginal and queens suppress worker reproduction by physical aggression. Highly eusocial species on the other hand, have large colonies, queen and workers and sometimes even different groups of workers morphologically differentiated, caste differentiation in preimaginal and queens suppress worker reproduction. Colonies of most ants and termites include workers of two or more sizes. Workers of a particular size specialize in tasks, smaller ant workers labour in the nest, while bigger individuals defend and forage. Morphological polymorphism relating to allometric growth of jaws in castes specialised for defence is known in termites and ants and not is bee and wasp societies.

Fig. 5.6. *Oecophylla smargadina* (after R. Gadagkar)

Fig. 5.7. *Diacamma ceylonense* (after R. Gadagkar)

Eusocial insects are among the most dominant groups in the history of life comprising a large part of the terrestrial biomass, channeling much of their energy flow and exercising a profound effect on ecosystems and the evolution of terrestrial organisms (Wilson, 2005). While ants are subject to colony level selection, wasps are subject to individual level selection achieving *social homeostasis*. Major workers in an ant colony tend to show for labour groups defending nest entrance and these are distinguished from major workers in the interior of the nest storing liquid food in their crops. ``Population of colonies tend to change in density, population stability through *social homeostasis*, and have the additive effect of permitting entire fauna of social insects to survive in habitats with severe irregularly fluctuating environments (Wilson, 2005). Wilson calls the insect colony as a `super organism' involving mass communication and adaptive demographs by which the behavioural repertoires of individual workers are integrated". Eusociality has been explained on the bases of Hamilton's (1964) `inclusive fitness theory', which maintains that polyandry or multiple mating by queen and polygyny, the presence of more than one egg layer in the colony tend to lower intracolony genetic relatedness. Opinions tend to differ in regard to mutualism being important in the early stages of social evolution and kin selection might have acted in later stage (Gadagkar, 1992).

Ants as balances in nature

Veeresh (1999) has highlighted interesting examples of the utilitarian values of ants as balancers in nature for pest problems, benefitting man by contributing to pollination, soil improvement, nutrient cycling, and as biocontrol agents. The oldest known record of the use of ants in biological control of insect pests is *Oecophylla smargadina,* an arboreal nesting species buildiing nests by webbing broad leaves and useful in controlling coreids, mirids and lepidopteran larvae in plantations. In *Oecophylla* there are several hundreds of workers in a main colony in trees and about 50-60 workers in a subcolony. Subcolonies are formed mostly broad-leaved plants or small trees like mulberry, *Pongamia, Erythrina, Morinda* etc, while the primary nest occurs in fallen trees like Mango.

Team work involved structured cooperation is very evident in ants such as *Oecophylla.* to accomplish the act of gluing leaves together, the team work involved relates to (a) initially groups of individual must pull leaves together and hold them in place (b) clue-like silk produced by the larvae (c) a group of workers must hold the larvae and use their like a puler of glue, to bond leaves together, followed by gluing. In short, such team work adds to superefficiency. This is a structured coordination.

The primary colongy has the reproductive queen and male ants, with a large number of workers which form trails connecting the primary nest with subcolonies on other trees. The queen stays in one nest of the colony and the workers carry grubs from the main to sub colonies to feed them by trophallaxis.

Species like *Solenopis geminata* collect seeds from a large number of weed species. Ant-plant mutualism is known in species of *Crematogaster, Camponotus, Monomorium, Tapinoma* and *Solenopsis*. The subfamilies Ponerinae and Dorylinae comprise many predatory ants with *Leptogenys processionalis* predatory exclusively on termites. As many as 25-35 thousand termites are taken every day by each colony of this ant, keeping the termites damage to plants at a minimum (Veeresh *et al.,* 1995). In minimizing termite attacks to forest trees, ponerine ants such as *Megaponera foetens* have been used. *Ant mosaics* are known, wherein a small number of dominant ant species are commonly found to maintain mutually exclusive territories. Besides these, a second complement of ant mosaics co-occurs with each of the dominant species, patterns of which are behaviourally controlled by different tolerance levels among ant species. The key role of homopteran honey dew and plant nectar, aid in the structure of ant communities. *Trophobiosis* or nourishing life, has proved one of the most successful aspects, with aphids, scales, mealy bugs, tree hoppers, caterpillars of lycaenid butterflies, offering sugary section to the ants for food and in return are protected from enemies. Ants sometimes take them inside their nest as virtual members of the colony and this symbiosm is trophobiosis (Holldobler & Wilson, 1995). Ant acacias are adapted for feeding as well as housing acacia ants. Swollen thorn acacias have extrafloral nectaries of the petiole. Ants feed on the sugary solution, secreted by these nectaries and in turn ants keep the plants free of insects and fungi. *Pseudomyrmex* ant colonies have over 30000 ants and spread across trees.

Interesting information is available relating to numerous species of seed ants and *Pogonomyrmex barbatus* a desert species, with a stable colony behaviour, with colonies living upto 10-15 years. There are patrollers which recruit other ants to retrieve food. Interaction among workers groups change as the colony gets older. Workers change tasks and non-foragers become foragers if needed, but foragers do not change tasks. It would not be out of place to make mention of the leaf cutter ants of the Neotropics. They are the dominant plant eaters of the region and 15% of the leaf production of the Tropical forests disappears down the nests of these ants. In the nest the leaves are shredded and inoculated with a particular fungus which digests them and is in turn consumed by the ants. They grow the fungus, a kind of mushroom in underground chambers about the size of a football. A single leaf cutter ant nest may contain a thousand such chambers, upto 18 feet and support a society of more than a million ants.

Mass communication occurs involving the transfer of information among groups of individuals of a kind and the flow of food in colonies is regulated by communication between groups.

Fig. 5.8. *Polistes stigma* (after R. Gadagkar)

Gordon (1999) in his chapter on `the myth of the ant queen' indicates as `to how ants colonies develop a mystery that has implications that extend far beyond our brains and immune system. Her work focusses on the connection between microbehaviour of individual ants and overall behaviour of colonies" An interesting observation regarding the queen: She lays eggs and are fed and cared for by the workers. In a harvester ant colony, with many feet of intricate tunnels and chambers and thousands of ants, separate the queen, surrounded by interior workers from the ants working outside the nest with only the chambers nearer the surface. It is virtually impossible for the queen to direct every worker's decision about which task to perform and when. The harvester ants that carry the queen off to her escape hatch, do not do so because they have been ordered by their leader; they do it because of the queen and is responsible for giving birth to all members of the colony as it is in the colony's best interest and the colony's gone pool, to keep the queen safe. The genes instruct them to protect their mother, the same way the genera instruct them to forage for food. In other words, the matriarch does not frame her servants to protect her, evolution does". The social cohesion of ant colonies is held up to their generations, the suiter ants popularly a colony sharing there quentum of the gives due to a complicated process of sex-determination in ant socities, the spared genes implying a greater communal intact in preservation (Hamilton, 1964).

Fig. 5.9. *Vespa affinis* (after R. Gadgakar)

A group of ants called honey-pot ants are typical of Australia. They leave the nest, but inhabit underground fallen trees six feet down. Foraging workers with crops full of honey and nectar feed it which swells its abdomen and cling to be roof of the fallen tree and hundreds hang in rows. During the dry season, the active workers visit the underground honey pots.

The leaf cutter ant of South America is one of the most complex of societies, with the rest of a colony, gigantic. Below ground it reaches as much on 18 feet in a maze of chambers nad pathways. A processing of ants carry a freshly cut segment of a leaf hoisted above their backs, like a flag. Soldiers guard the community, with heads swollen is accomodate their huge jaws. These ants by multiplying and organizing into a society, have become masters.

Sociality in Thrips

Thrips comprise an order besides the Hymenoptera where females are diploid and males haploid. The interplay between haploidy and kin selection in social evolution makes thrips useful insects in the study of sociality, particularly in view of haplo-diploidy being a key factor in shaping the evolution of sociality in the Hymenoptera. Hamilton (1964), Trivers and Hare (1976) and Kranz *et al.* (1999). Six species of Tubuliferan thrips induce galls on Australian acacias and are known to be eusocial (Crespi, 1992, Crespi and Mound, 1997). These have a morphologically distinct social caste with reduced wings, antennae and enlarged forelgs that fight invades. They are of interest in the evolution of social behaviour, because their life history differs markedly from the other social insects. For instance, thrips do not forage or nurse their young as do social Hymenoptera and this is because galls are nutritional sinks wherein the insects feed on the cell contents of the gall ensuring a continuous supply of nutrients. Thrips of the genera *Oncothrips, Kladothrips, Onychothrips,* with 21 species form galls. Species with a `soldier caste' however belong to the genera *Kladothrips* and *Oncothrips* their galls being inhabited by a single female the first brood consisting entirectly of soldiers. This brood is followed by a second dispersing generation in which adults are not morphologically distinguishable from the gall foundress (Kranz *et al.*, 1999). The foundress could be influencing the sex-ratio of the soldier's brood, by producing male soldiers first. Protandry in males maximises the possibility that female soldiers are mated as soon as they are mature. The foundress could be influencing the sex ratio of the soldiers brood by producing males first. More than half the soldiers are almost exclusively laying fertilized eggs.

Mound and Moritz (2000) distinguished six suites of species associated with phyllodinous acacia trees. Members of one genus form a suite of *squatter species;* the gall induces a suite with 21 species; the *domicile-constructor* suite which glue new phyllodes together; Besides are the *kleptroparasite* suite, *the inquiline* suite and *opportunistic* suite, with species occupying abandoned galls. Females of some gall thrips develop extreme *physogastry* laying hundreds of eggs and the high fecundity tending to attain reproductive division of labour.

Functional groups of forest litter ants are evident, with such guilds as omnivores, general predators, termite predators, collembolan predators, ants preferring predators, seed eaters and specialist predators, reflecting, the influence of forest vegetation in structuring ant community composition (Vineesh *et al.,* 2007).

Eusocial insects are ecologically dominant,with members of a community in frequent communication, segregating according to caste and to the local condition each one encounters around the nest. Social insect colonies are able to maintain more nearly constants nest environments than solitary insects. As such social homeostais has evolved to control virtually every conceivable component of the environment. The unit of natural selection in the colony and the age and size frequencies of the colony member are adaptive traits (Wilson, 1990). Mass communication integrates the behavioural repertoires of individual workers, often employed during recruitment.

Being opportunistic, thrips species are known to switch between phytophagy and predation, between sporophagy and predation, between pollen-feeding, leaf feeding and predation and between completly unelated host plants of different localities (Mound *et al.*, 1996). Further, in mycophagous species such as *Elaphrothrips* which live in groups in dead leaves, large males have been observed each protected a female with a groupof eggs and driving off or fighting with rival males. Those large males have enlarged forfemora and exceptionally large foretarsal tooth with which they stab a rival female during combat. Sneak mating is observed in gynaecoid males with more slender forefemora and reduced foretarsal tooth. Similarly many often are spore feeders like *Tiarothrips, Hoplothrips*, show similar behaviour. Interestingly enough the males of the giant thrips *Idolothrips spectrum* from Australia, living in dead, hanging encalyptus leaves have a different method of fighting, and since they lack foretarsal teeth, they use their abdomen as a weapon which beams enlarged lateral abdominal tubercles. Mating strategies and fighting methods tend to vary within and between populations. As has been indiçated, the larger males may be involved in parental care through the defence of egg masses from natural enemies. (Mound, 1991, Crespi, 1986).

Polymorphism and structural allometry are associated with competition for resources, including male/male combat resulting in diverse ecological and behavioural aspects, adults leading juvenile groups to particular feedings sites along pheromonal trails and leading them back again. Extreme foreleg allometry and strong armature, especially more in apterous forms are indicative of the fighting abilities, the foreleg armature being mostly associated with wing polymorphism, the apterous males exhibiting diverse structural pattern in the different species. Variation in fecundity enables evolution of reproductive division of labour, while in males the extreme development of foreleg armature or lateral abdominal horns

or thoracic processes enables defence against invaders - aspects significant in the species of *Ecacanthothrips, Bactrothrips, Dinothrips, Elaphrothrips, Tiarothrips, Mecynothrips, Hoplothrips,* to mention the major genera. Increased variations in fecundity tend to favour the reproductive division of labour with such forms as oviparous, ovoviviparous and viviparous individuals with corresponding changes in the ovarian structure. Fighting behaviour in males is equally characteristic of polymorphic males, with egg guarding and larval defence behaviour being typical of these species.The number of eggs laid varies with genera and species from less than 50 to over 300 in a single groups the adults guarding the eggs. In some cases predatory ants are effectively repelled by a secretion containing a gamma delactone. Females of some species of *Elaphrothrips,* where maternal care is known, use a potent defensive secretion, juglone to repel predators such as salticid spiders.

Developmental times also tend to vary with species, being 15-25 days in *Elaphrothrips,* 20-25days in *Tiarothrips* and 20-40 days in *Bactrothrips.* Oocytes developed within lateral oviducts up to a fairly active stage of development in ovoviviparous females, the eggs hatching a few hours after laying; while in the partial ovoviviparous females, the oocytes developed in the lateral oviducts up to germ-band formation, with subsequent egg-laying and hatching occurring one or two days. Patterns of seasonality, habitat stability and population structure appear to account for the evolution of advanced parasociality in many species. Both genetic factors and extreme ecological factors appear important in the evolution of such sociality. Communal resting sites have been identified when not foraging, such sites occurring in depressions or cracks on the bark. It is interesting to note while foraging, that adults keep the apex of the abdomen upward, waving them. In many species, the males are territorial at the oviposition sites.

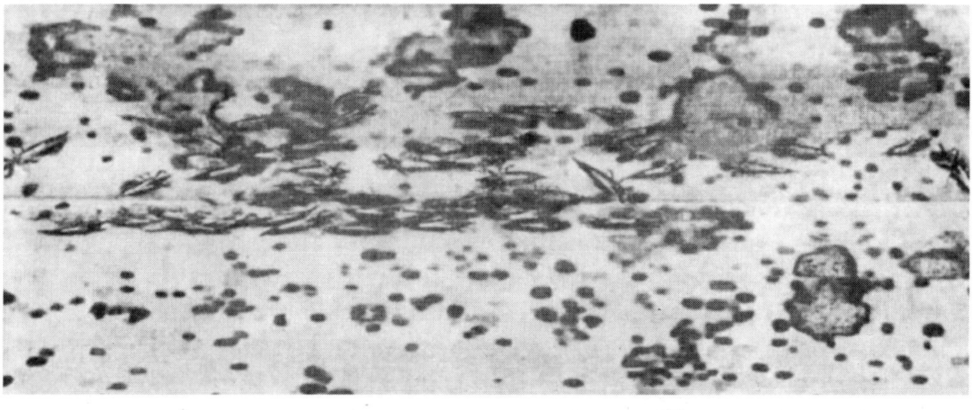

Fig. 5.10. Mycophagous thrips on dead palmyra leaves.

Tiarothrips subramanii accounts for 300-400 individuals in a colony and the escalated fights by males, challenging of extreme males by subordinate males, and slabbing and wagging behaviour. Armed extreme males are associated with territorial defence of oviposition areas, the dominant males securing 80% of the matings. Intraspecific ecological succession with macropterous forms succeeding the apterous form is evident in some species of *Hoplothrips*. Brachypterous forms are produced in some species when the fungi are fresh and vigorously growing, while the macropterae are produced as they age and deteriorate, so that there is intraspecific ecological succession, the macropterous forms succeeding the apterous ones. An extreme degree of aggregation and sexual selection is evident in sporophagous idolothripines, the nature of aggregation and the population in a colony greatly depending on fungal food availability. The larger size in males appears to confer greater probabilities of success in the reproductive investment of the thrips, and the behaviour of the female to choose a large male will be favoured and maintained. The temporal and spatial structure combined with the mode of development have led to the evolution of social behaviour and the highly patchy habitat provided by dead trees tends to be conductive to the evolution of eusociality with inter-generational stability of feeding sites. Patchy, but inter-generationally stable food supply is the primary factor, which is also the reason for the promotion of social behaviour. Colony formed by multiple adults or pleometrosis typical of wasps, bees and ants, is equally evident in these species, since males develop by arrhenotokous haploid parthenogenesis from unfertilized eggs, while females develop from fertilized eggs. It is predicted that females tend to reproduce by oviparity when there is local mating competition, and by ovoviparity when constrained by predators with or without local resource competition. Reproductive females which are involved in local resource competition for food and oviposition sites select ovoviviparity as they can retain eggs till a suitable rearing site is located.

Therefore, an indepth assessment of aggregation of species in relation to their reproductive output, behavioural diversity of gynaceoid and oedymerous males, egg size and reproductive division of labour contributes to a better understanding, of eusociality. Phenotypic changes alter fitness of the genotype in termsof oviposition mediated through mating patterns. Fecundity index expressed as the number of eggs, female as in oviparous forms and number of larvae/ female in viviparous forms, is an added measure of sociality. The impact of genetic, phenotypic and ecological causes relating to the structural diversity evident in apterous males as against macropterous ones, is an aspect deserving further scrutiny in relation to better appreciation of cooperative endeavour evidenced in mycophagous colonies of thrips. The genetic relatedness of the diverse morphs, i.e. egg diversity, clutch size and output or larger eggs, besides competition by females for food and space tends to increase competition. The relationship between body size, armature and reproductive success deserves further scrutiny. According to Wilson (2000) forms with solitary lifestyle require alteration in every system of the body, coupled with sufficient plasticity in the traits prescribed by genes among adults castes. As has been indicated by Hamilton `the coincidence of male winglessness, female-based sex ratios, male haploidy and male fighting in spatially structured populations is probably not accidental'. Further studies are needed in relation to diverse mycophagous species to be able to appreciate the diversities expressed and needless to emphasize that patterns of sociality, habitat stability and social structure tend to account for sociality in thrips.

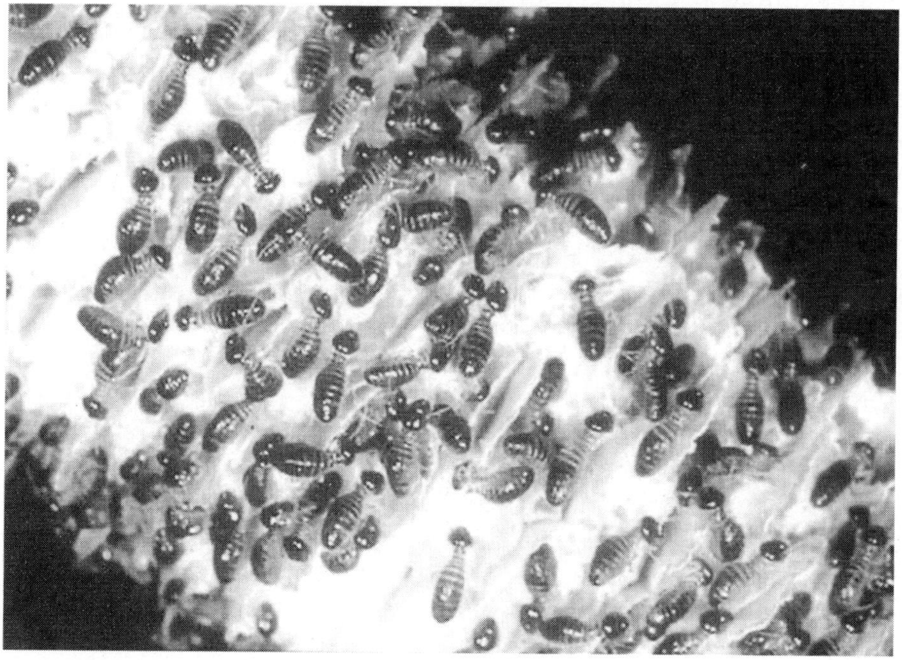

Fig. 5.11. Termite workers (Courtesy V.V. Ramamurthi).

Wilson (1990) indicates that termites are the "master-brokers of the decomposition of dead plant material" and both termites and ants are extraordinarily abundant. To quote Urbani *et al.*, (1978). ``130,000 termites and 16.6 million ants per hectare in Tropical forest litter and 1.8 million to 180 million termites per hectare in several localities". Both ants and termites have been designated as ``superpowers of the insect world. More than 8000 ant species and over 2000 species of termites are known. If you examine an earth mound, there will be a large central cell occupied by the egg-laying machine, the queen, with an abdomen enormously enlarged, filled with eggs and a much smaller king. Secondary queens also develop in adjacent cells. Workers and soldiers attend to nest building, care of the young, food production and distribution. Soldiers defend the colony and are of two kinds, the *mandibulate* and *nasute* types, the former with strong jews, while the former have minute jaws with a gland at the tip of the snout, which secretes a sticky, poisonous fluid. This fluid is ejected as a fine jet on the enemy, immobilizing its. In a well developed community two other types of reproductive individuals occur-those with rudimentary wings and with reduced reproductive structure and wingless form with further reduced reproductive structure. Termites feeding on wood, have to digest cellulose and for this they depend on protozoa called *Trichonympha* which they harbour in their intestine. The trichonymphids break-up the cellulose and subject the food to the action of the digestive of juices of the termites. The reproductive individuals are fed with the regurgitated digested food by the workers. The trichonyphids occur only in the intestine of workers and soldiers. Some of the common wood developing termites are *Kalotermes, Neotermes* and *Cryptotermes*. The ground dwelling termites include the mound building ones, *Macrotermes, Microtermes, Odontotermes* and those that construct nests underground, and the *Coptotermes* and *Heterotermes*.

Unlike termites, the honey bees live in combs which they build. Three types of individuals occur in a community, the males or *drones,* the female *queen* and sterile workers which are undeveloped females. All are winged. The queen is the mother of the colony, with only one queen at a time whose function is egg-laying. The drones fertilize the females and take no part in the upkeep of the colony. The workers build combs, collect pollen and nectar from flowers, keep the colony cool and clean. The workers have flat, smooth tibiae which are groomed and fringed with hair, serving a pollen baskets for storing pollen carried to the comb. The workers have a reservoir for collecting nectar from flowers, into crop. This nectar is regurgitated from the crop into the cells of the comb and this special kind of food is fed to larvae which are to become the queen and drone, is called the royal jelly. There are distinct types of cells for the development of the workers, males and females.

There are three species of honey bees – *Apis dorsata, Apis florea* and *A. indica*. *Apis dorsata* in the rock-bee which is found in forests building single comb on tall trees and along the ridis of steep rock. Because of its large size and ferocious temper and nomadic habits, it cannot be domesticated. *Apis florea* is the litter honey bee which builds comb on the branches of trees and *Apis indica* builds combs in parallel rows in the hollows of trees, rocks and other protected spots. This can be domesticated and reared in apiaries.

Sociality in Caterpillars

Many caterpillars show patterns of integrated behaviour, several being gregarious as larvae, with caterpillar of butterflies better known then moths and achieve numerical abundance/dominance in the communities they inhabit. They are ``obligatory gregarious species" with a foraging behaviour and behavioural patterns are associated with chemical communication. They show sociality in feeding, cooperative shelter building, thermoregulation and antipredator defence. The tent caterpillar is the best known of social caterpillars. Aggregation of social caterpillars vary in size from a few individuals to assemblages of hundreds or even thousands of larvae as in *Eupterote mollifera.* Many females choose oviposition sites close to egg masses, forming multicolony aggregations. Survivorship is influenced by colony size, ability of caterpillars to locate sufficient food, defense against predation, constant shelter and maintainance of body temperature for growth and development (Stamps, 1981).

Fig. 5.12. *Eupterote mollifera* Caterpillars (Courtesy: Prasanth Jacob)

Caterpillars that forage together may aggregate as a community build silk or leaf sheaths, creating a microclimate conducive to growth. *Malacosoma* is the best known example of tent caterpillar, besides *Yponomeuta* or web worm. Leaf shelter builders harness forces generated by stretching silk threads before attaching to leaves. Such shelters faciliate thermoregulation and predator avoidance. As regards foraging, tent caterpillars lay down chemical trail markers from the tip of the abdomen, moving to previously unmarked branches in search of food. The trail based foraging strategy of chemical communication facilitates group foraging.

Fig. 5.13. Top. Sciarid fly larvae, Bottom: Tent caterpillar (Courtesy Douglas Whitman)

All social caterpillars appear to have been derived from egg clusters. ``Cluster oviposition arise first and sets the stage for the evolution of larval adaptation that could lead initially to the maintainance of contact and eventually to sociality" (Fitzgerald, 1981).

CHAPTER - VI

Cecidogenous insects : Diversity, biological and ecological interactions

Gall formation or cecidogenesis is a complex phenomenon involving the recanalization and reorientation of plant development resulting in the concerned insects becoming partially or complexity enclosed, enabling gall insects to grow, mature and reproduce within the galls. In short, galls represent the functional efficiency of the host plants as an outcome of insect feeding or oviposition. Galls represent a biological system evoked by insects, the system being a complex one involving the interactionof nine than one organism, including the gall maker inquilines and predators. Such a biological system involves a highly specific organized community with three-level interactions among the participants. The highly specific nature of the gall insects to particular hosts throws sufficient light on their behavioural patterns of colonizing specific areas. All the same, the interplay of two unrelated organisms as in gall systems highlight the evolution of a remarkably efficient interacting system, the plant mobilizing all the resources to protect itself initially and subsequently undergoing a series of adaptive changes to enable the survival and sustainance of insects (Raman, 1984). The continued presence and stimulus of insects is essential for the development of specific gall form and structure. Insect stimulation tends to redirect the host plants cellular and organismic activities.

Galls represent discrete microhabitats that support relatively closed communities of specialised inhabitants. They range in complexity from relatively open depression, folds or curls to structures in whiich the galler in inclosed entirely by host plant tissues. Multiple, highly differentiated layers of plant tissue are characteristic of more complex gath. Gallers can manipulate the allocations of plant resources to their advantage, through their ability to concentrate plant nutrients and metabolities in gall tissues, both by elevating photosynthetic ratio in affected parts and by mobilizing resources from neighbouring plant tissues.

Fig. 6.1. Petiolar galls of thrips

Several insects such as thrips, aphids, coccids, cecidomyiids, cynipids, to mention a few, are known to alter the developmented profiles of plants, manipulate plant tissue to form galls varying in form and structure in accordance with the concerned is a major gall induction in a major specialised type of herbivory and galls are miniecosystems or plant species and the insect galler food webs. In these interactions the adaptive nature and ecological significance are evident in the nutritional and structural diversity, the gall maker having precise control over gall differentiation and development. Most galls involve inhibition of development, differentiation, growth and tissue suppression alongside with the activation of some tissues. While hymenopteran and dipteran galls dominate the scenario, equally fasicinating gall diversity is evident in the galls formed by thrips, aphids, mites and coccids. Factors affecting gall size, include the number of larvae, diversity of galls, portion of infested leaf etc. A high degree of host specificity is typical of galls, the plant tissues responding to the stimulus, when they are juvenile, meristematic and differentiated. The size, structure and metabolism of galls are under the control of the gallers which may range from an individual to many.

Fig. 6.2. *Mangalorea hopeae* on *Hopea ponga* (after Raman)

Adaptive diversity

The developmental processes of the host are manipulated to produce structures enably feeding, shelter and escape. The extent to which galls manipulate growth, varies considerably from induction of cell proliferation to the production of complex structure. Variation in the gall phenotype influence the fitness of insects and gall phenotypic variations are also affected by genetic variation in insect populations. Gall formation represents a very precise and species specific manipulation of host plant chemistry, involving a reduction or increase in critical nutrients within the gall to improve insect performance. Cecidogenous insects select parts of the plant that have high levels of nutrients value and the nutritive tissue acts as a since wherein all the important metabolites enable growth and development of the insect. Gall formation as such represents a very precise and species specific manipulation of host plant chemistry, the gall environment favouring insects with improved nutrition, reduced defence chemicals and limited access to natural enemies.

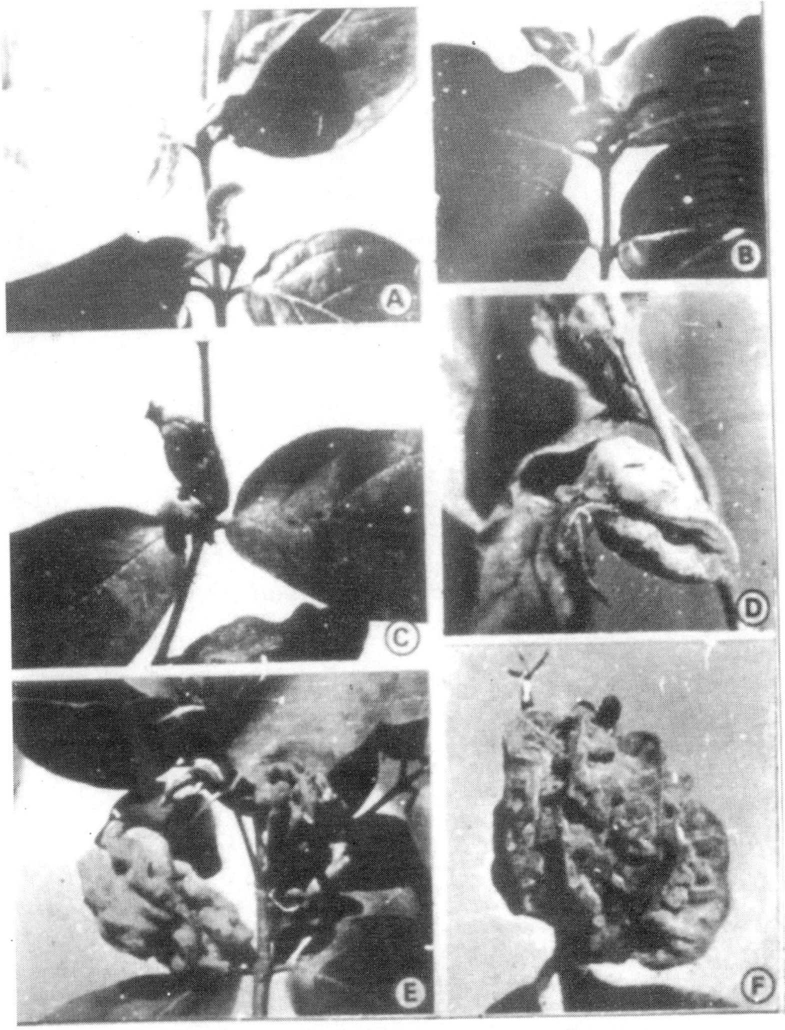

Fig. 6.3. Stages of development of *Calycopterus floribundus* by *Austrothrips cochinchinensis*

Variation of gall phenotype influence the fitness of insect and gall phenotypic variations are affected by genetic variation in insect populations. In short, galls are harmonizously organized entities with an orderly arrangement of cell layers and determinate growth which result in structures with peculiar size, shape and the diversity of gall form is very typical of many thrips galls. The number of generations that occur within a gall is variable and among many examples, that of the woody galls of Casuarina as well as of *Calycopteris floribundus* pounch galls, containing thousands of thrips, are outstanding.

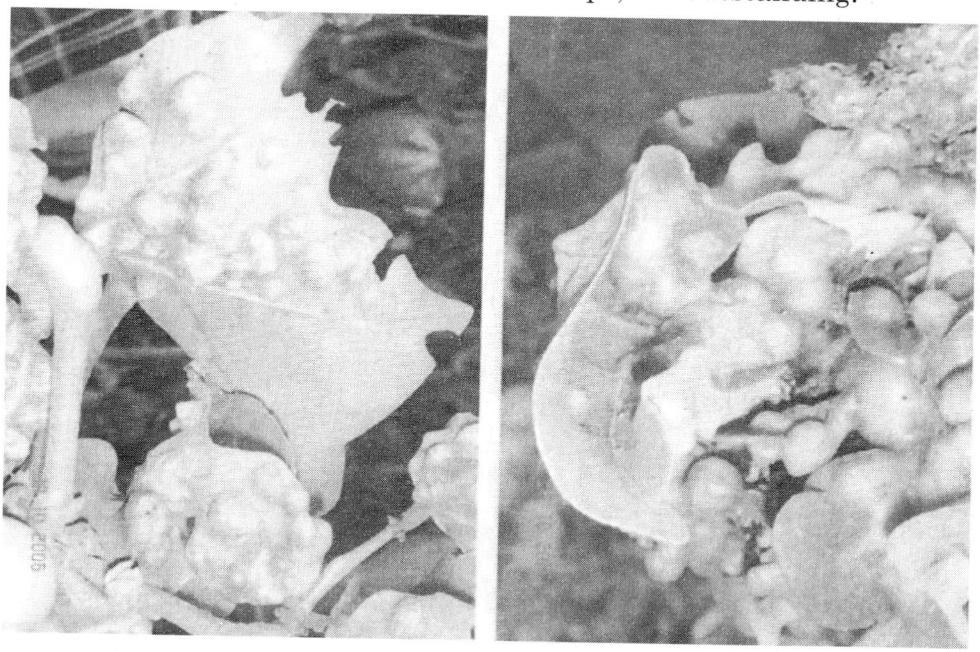

Fig. 6.4. Erythrina galls formed by *Quadstrichus erythrinae* (Courtesy, Prasanth Jacob)

Diversity of galling in relation to the impact of galls communities

Gall thrips and Homopteran gall inducers like psyllids, aphids, have numerous gall-inducing representatives with gall thrips abundant in tropical host plants. New leaf flush in the host plant is one of the primary criteria that decides population build-up within galls. Rarely species such as *Thilakothrips babuli* inducing galls on the buds of *Acacia leucophloea* show seasonal incidence with the lifespan spread over 40-45 days; when the same gall thrips occurs on the inflorescence the life span is much shorter, being 20-25 days. Gall forming trait is dominant among thrips with more than 400 species on diverse angiosperm families. While in most of the galls the number varies from 30-100, species such as *Austrothrips cochinchinensis* inhabiting large punch galls of *Calycopteris* harbours nearly 10000 individuals.

Gall forming aphids are confind to the Eriosomatidae, Adelgidae and Phylloxeridae, while psyllids with about 2000 species induce galls on a variety of dicost hosts. Several families of Coccids, Pseudococcidae, Eriococcidae, Kermisidae, Margorididae, Asterolecanidae, Diaspididae, include several gall inducing species and exceptionally males and females of coccids such as *Apiomorpha* induce seperate disimilar galls known as andro-gynocecidia in Eucalyptus (Gullon, 1984). Gall wasps and fig wasps have a narrow plant spectrum, but include a large number of species. Saw flies or Tenthredinae, eulophids cynipids, agaonids, eurytomids and chalcids are gall formers, on diverse hosts. Recently an eulophid *Leptocybe invasa,* an invasive species from Australia has affected Eucalyptus plantations in South India, besides a gall wasp *Quadrastichus erythrinae* another invasive species which has done considerable damage to *Erythrina.* Agaonids are true gall insects developing in the ovary of gall formers, present in fig syconia. Among the Diptera, cecidomyiids are the best known with around 2000 gall flowers, with 70% belonging to *Contarinia* and 60% to *Asphondylia* (gall midges have one or two or several generations a year). Cynipids among the Hymenptera such as *Aphondylia* and *Lasioptera* are very common. Different groups belonging to the family Cynipinae include the gall wasps which display a very complicated life cycle. Besides arrhenotokous parthenogenesis, they also show thely tokons parthenogensis as well. Alternations of generations involving a sexual generation of males of female and an agamic generation of only females is very common.

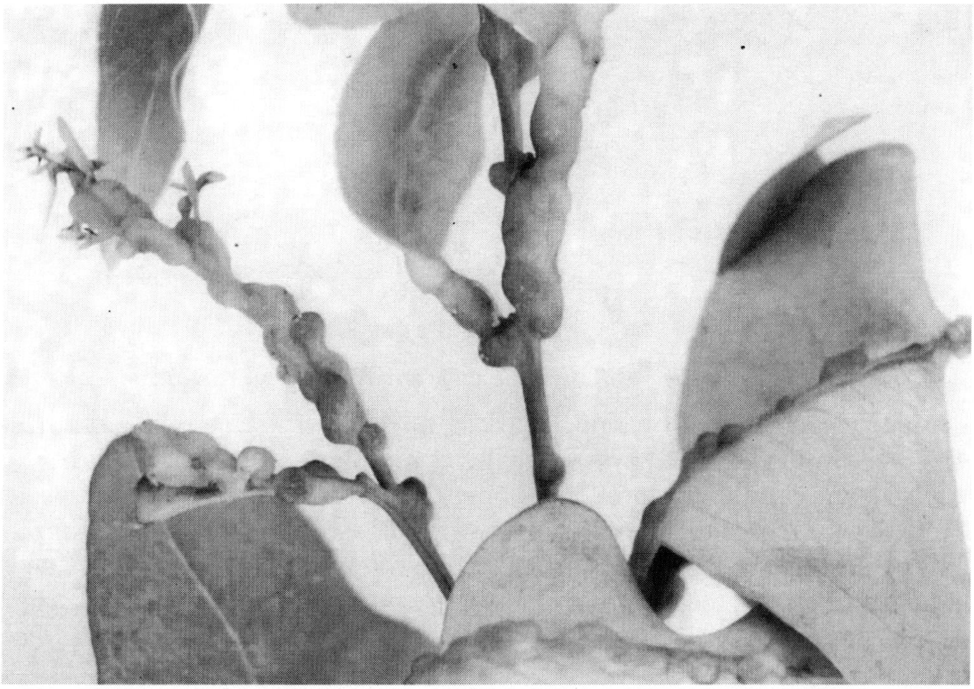

Fig. 6.5. *Eucalyptus galls* caused by *Leptocybe invasa* (after Prasanth Jacob)

While only a brief outline of the diversity of gall groups are indicated, the majority of the gall formers are highly specific to their host plants and the relationship between the galling insects and their host plants exemplify a special biotic relationship. The diversity of form and structure and population interactions within galls in unique. Several exquisitely shaped galls occur in diverse plant families with equally diverse population densities, which are critical factors that influence the size, surface thickness of the gall, with gall populations displaying an enormous range of variability. One cannot overlook the annual gall patterns observed in the coccid gall *Mangalorea hopeae* on *Hopea ponga* (Dipterocarpaceae) and *Calycopteris floribundus*.

Fig. 6.6. Eucalyptus galls by *Leptocybe invasa* (after Prasanth Jacob)

Successful exploitation of a host plant by a gall maker requires that an adult female first recognize and accept a given plant for oviposition. Once accepted the host plant development of must be suitable for the development of the offspring. In aphids for example, large leaves are adaptive, the asexual function increasing with leafsize, mothers often choosing to settle on smaller leaves rather than share, a larger leaf, since resource level could be less on a crowded leaf than an uncrowded leaf. The selective advantage of preference for a superior host plant could be density dependence (Abraham & Weis,

1991). Hence prefence and performance relationships are influenced by the breadth of the feeding niche, competition and natural enemies. A preferred oviposition site is attacked at a higher rate and performance, while performance include survival of immature stages, larval growth efficacy and resultant fecundity.

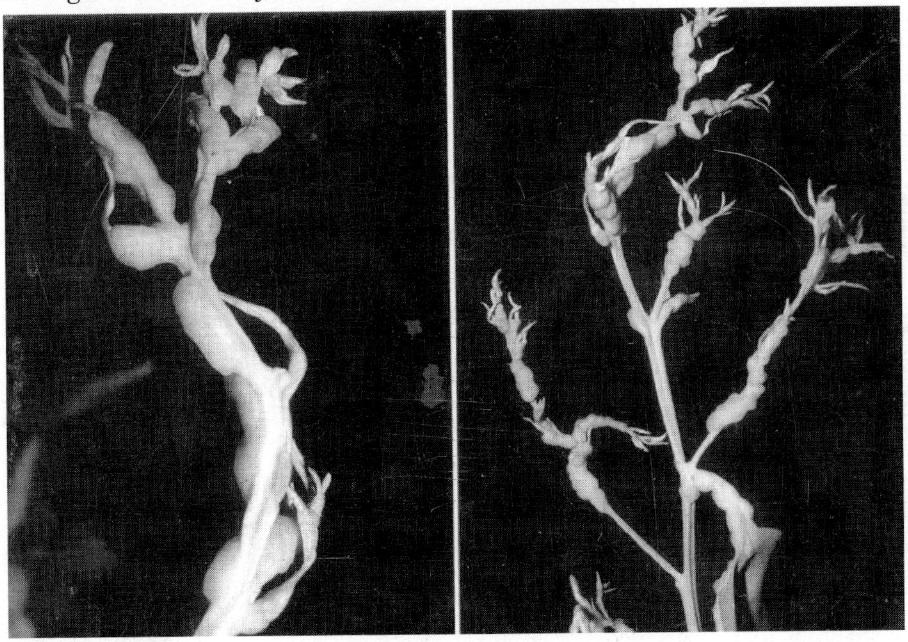

Fig. 6.7. *Leptocybe invasa* galls in Eucalyptus (after Prasanth Jacob)

Polymorphism and Social behaviour

Ecological opportunities for gall inducing cecidomyiids in different species of Anacardiacae include this capability of inducing galls on mostly leaves and flower stem occasionally. The Mangifera gall inducing cecidomyiid interaction offers fascinating insights with gall-inducing insect-flowering plant interactions.

The occurrence of polymorphism with major and minor females and ocdymerous and gynaecoid males, very typical of thrips, enables intra and interspecific competition regulation within galls. The reproductory behaviour of the polymorphs considerably influences fecundity and also result in patterns which lead to recognizable variation in fecundity and subsequent production of morphs. A further adaptive strategy relates to the ability of some gall thrips to switch on to the formation of a duel gall-leaf and inflorescence galls on the same host, as exemplified by *Thilakothrips babuli* on *Acacia leucophloea*, producing rosette galls on leaves and floral galls. The life cycle is shorter within infloresence galls and longer in rosette galls.

The recognition of the *open gall systems* such as induced by thrips where maturation and development of galls are largely dependent on the build-up of populations, and *closed gall systems* where cecidogenesis is controlled by the action of a single gall maker as in cecidomyiid and cyniipid galls, with an intermediary system in homopteran galls wherein some galls are induced by a population and other by single organism.

The biodiversity and biogeographical implication in species richness of gall inducing insects have been discussed in an exemplary way by Raman (2007) ``Patterns of radiation among gall-inducing thrips appear more complex than those that are known in gall inducing tephritids because of complex host plant relationships''. The best examples of galls on monocotyledons are the tubular leaf sheath galls induced by *Orseolia oryzae.* Ecological opportunities for gall inducing cecidomyiids on different species of Anacardiaceae include this capability to induce galls on leaves mostly, and occasionally on flowers and stems. A close understanding of the patterns of distribution and radiation of gall insects is a necessity for a better appreciation of galling system not only in cecidomyiids, but also in other groups of gall forming insects. As Raman (1984) has rightly opined, the interplay of two unrelated organisms as in insect gall system highlight the evolution of a remarkably efficient interacting system, the plants mobilizing all possible resource to protect itself initially and then undergoing a series of adaptive change to enable the survival and sustainance of insect and insect on the part displaying feeding specialisation to create a new habitat. The sequence of events envisaged in development of a cynipid gall community, parasite colonization, radiation, and specialization and such gall communities have initially polyphagous parasites and only after a lengthy association with a host, do some of these, as a result of interspecific competition become specific parasites. New polyphagous parasites are being continually added to communities, replacing some of these previously established (Askew, 1984).

Diversity in the exploitation of plant tissues leading to gall development being typical of galling insects, the astonishing diversity of gall shapes and sizes population, and structural diversities resulting in abnormal differentiation of tissues and the accompanying cytological and physiological changes in host plant tissues enable adequate nutrition to the galling insect. The continued presence and stimulus provided by the insect, though basic for gall development, adaptive strategies involve hyperplasy and hypertrophy, besides meristematic capability. The continued presence and stimulus of the insect is the most crucial element for the development of a specific gall form and structure. New tissue types are generated by hyperplasy, besides their ability to revert to a meristematic state followed by rapid cell division. Efficient resource partitioning and effective utilization of the host ae typical traits of gall insects, thereby maintaining a special biotic relationship, with the gall insects spending a major part of their life cycle on the host plant, within the protected environment. The inductive behaviour of the insect and the responsive behaviour of the host plant regulate the degree of interaction between the two. While gall inducers are strikingly monophagous, rarely do we find them accepting novel hosts within the species of the same genus and such host shift processes are not uncommon, besides the ability of two or more species to form different kinds of the gall on the same host. A gall therefore, is a complex `biological expression' involving the gall insects and the plants - an outcome of the host plant's effort to isolate the insects in space and time. In view of the localized nature of cell divisions at unconventional sites and active growth, gall formation essentially involves differential growth patterns. To quote the Cecidologist, Mani ``In so far as the reactions of a plant favouring the survival of the organ attacked by the gall-inducing organism, the primary advantage in the formation of the gall is to the plant. By producing the gall, the plant has in a sense localized the parasite in special tissues and has forced it to extreme specialization". The view that the galls are of benefit to both plants and insects is equally prevalent with a perfect co-ordination between the gall inducer and the gall bearing plant. Rohfritsch and Shorthouse (1992) opine that the cytological and biochemical aspects of the nutritive tissues lining the gall lumen appear to favour the insect. Insects control the redirection of growth and physiology of the attacked tissues to the insects' advantage. Diverse and complicated patterns and mechanisms of colonization are known, resulting in `adaptive radiation' and it is believed that gall-inducers are in a ``dynamic evolutionary state" with different gall types and galling sites on their host plants. Interestingly in some midge galls, an additional participant is fungus, resulting in ambrosial galls, with their chambers covered with a layer of fungal hyphae on which they feed. Based on the different degrees of involvement of diverse insects and growth patterns, different kinds of galls result such as outgrowths, swellings, folds, rolls, pouches not to mention of the complexities observed in hymenopteran, dipteran and thrips galls. A chemical stimulus from the insect accelerates the velocity of cell division at the potential gall site on the host plant, the stimulus initiating novel patterns of differentiation at the attacked host plant site. In some, like gall wasps and sawflies oviposition stimulates gall initiation, while in others like gall midges, aphids, thrips, psyllids, feeding action and salivary secretions of the larvae stimulate gall development. The final shape of the gall is achieved only with the continuous and cumulative feeding impact of the offspring of the gall-forming female. One of the finest examples relate to the large pouch galls on *Calycopteris floribundus* induced by *Austrothrips cochinchinensis* nearly 8-9 cm long and 40-50 cm wide, harbouring more than five to ten thousand thrips in all stages of development. As against this large number, gall induction in phyllodinous Acacia trees involves a single thrips feeding on the upper surface of a young phyllode. The rate of increase of gall biomass is directly propotional to the feeding intensity of the inducer. The population of the gall maker has a decisive influence on the morphology of the galls. Four basic stages of gall development therefore involves initiation, growth and differentiation, maturation and dehiscence, the biomass of the gall becoming increased by cellular hyperplasy and hypertrophy.

While survivorship, development rate and functionality contribute to fitness, the choice of the gall site becomes important in view of the need to persist for a longer time till a new generation emerges. As such a high level of synchrony between the host and the gall maker phenologies become important. Some have a heterogenic life cycle like cynipids including a sexual and an asexual generation and the two produce very distinct galls, the emergence dates of the two generations being well timed. In short, the ecological and taxonomic criteria as well as species richness of gall insects on host plants are determined by several interrelated factors, which function differently in different groups.

Nutritive tissues and biochemical interactions

Gall insects exert a mobilizing effect on their hosts, effecting a relocation of nutrients and in so doing ensure a continued supply of nutrients. The galler should have the necessary mechanism to provoke the host to develop the nutritive tissue as well as the ability to alter their metabolic activity. A cytological gradient is evident in the nutritive tissues in relation to different zones and tehre is a progressive change from the edge of the larval chamber outwards. The first nutritive tissue layer shows nucleolar and nuclear hypertrophy with fragmentation and reduction of the vacuoles. The size of the vacuoles increases with a decrease of ribosome density in cells further away from the larval chamber. The nutritive cells lining the larval chamber have abundant cytoplasm rich in structural proteins and fat droplets with more numerous plastids and mitochondria. The nutrtive cells exhibit increased enzymatic activity with acid phosphatases and aminopeptidases well suited for larval feeding. A hyperactivity of peroxidases and polyphenol oxidases is evident throughout the gall tissues in all stages of development and the vigorous activity of these enzymes is seen in the nutritive zone cells and and beyond, suggesting a triggering of synthesis of these enzymes by the feeding stimulus of teh cecidozoan. The specialized structure and function of the nutritive tissue contribute considerably to the interactions, dynamics of the insects and the host plant, representing an altered cellular and metabolic environment. In short, these cells are cytologically dynamic and metabolically active. Hyperactivity of tannin is also evident in many gall tissues. As such specific phenological changes in their physiological attributes, contribute to effective exploitation of the host, the biochemical sustainability of which varies with age and is tissue specific. It is this synchronization of the life cycles with the phenology of the hosts that galls insects like thrips establish several generations. Several galling species display overlapping generations. The complex life histories of cynipids include overlapping generation within the pattern of alternation of sexual and asexual generation (Weis *et al.*, 1988).

Molecular and genetic basis of plant defense

An understanding of the molecular and genetic basis of plant defense and insect counter defense mechanisms become obligatory and such studies on gall insects should involve an understanding of the molecular strategies between them. Host plant resistance studies have indicated that plants possess many genes encoding defense related proteins. The genetic basis of interaction between insect biotype and host genotype, well reflected in the case of the rice gall midge *Orseolia oryzae* supports the gene for gene interaction. The rice gall midge larvae feed at the apical or lateral buds of the rice plant stimulating production of a covering gall. When fully formed it is a long, ivory-white tube that terminates in a solid plug of tissue termed the silver shoot (Fig. 1). Infestation triggers hypersensitive reaction (HR) in the plant, leading to extreme necrosis at the apical meristem in some gall midge resistant biotypes. Both HR^+ and HR^- types are observable, HR^+ being induced in response to rice plant infection, while HR- tends to be constitutive in nature. HR is a host induced response which invokes morphological and biochemical changes leading to destruction of the attacked tissue. HR leads to disruption of nutritive supplies to the invading organism and initiates production of toxic metabolites.

Increased instances of efficiency against insects are being idenfied in view of the postulation that HR (host plant resistance) is the most common plant resistance mechanism against gall forming insects. Polyphenolic substances tend to accumulate around the attacked cells soon after penetration by the galling larvae. In short hypersensitivity is a defensive response that encompasses morphological and histological changes terminaty in the death of the attacked tissue, results in the localization, containment, inactivation and eventual death of the insect (Yukawa, 2000).

Induced plant defense signaling pathways are now known with host plant producing signal molecules including jasmonic acid, ethylene, abscisic acid (ABA) and salicylic acid to regulate signal tranduction casacades in plant cells, leading to activities and modulation of defense related genes, with systemin, the 18 amino acid signaling molecule leading to the biosynthesis of jasmonic acid and defense proteinase inhibition. Gall midge-rice plant interactions have initially shown a compatible interaction with the insects forming the gall, as against the rice variety which is resistant to gall formation. Besides a secretion from the salivary glands implicated in gall formation, the involvement of plant growth regulators like Indoleacetic acid (IIA) are also known. With the gall intercepting, redirecting and prolonging activities, it provides a means to use host nutrients. While young gall cells accumulate nutrients, older galls synthesize phenolic and flavonoidic materials. Insects control the phenolic biosynthesis of their hosts.

Economic Cecidology

In recent years there have been increasing instances of gall formation in economically important plantation crops resulting in considerable economic losses. Of particular importance are the Asian Rice gall midge (*Orseolia oryzae)* causing serious damage to the rice crop. Diversity in pest population and rapid selections of virulent biotypes of this insect are aspects calling for new approaches. Determination of the economic threshold level for gall midge damage becomes important with 5% of the silver shoots and 10% of the tillers harbouring the immature larvae on the threshold level, warranting control. Alternate host of the rice gall midge are *Leersia hexandra, Echinochloa crusgalli* and wild rice, *Oryza nivara, O. perennis, O. rufipogon,* to mention a few. The life cycle of the rice gall midge varies under different temperature condition as for example, at temperature of 24-35°C, the deviation is 19-23 days, as against the winter months 16-32°C, then it is 25-34 days. The oviposition rate per female is on an average 125 yrs., with male : female sex-ratio of 1:3. One important factor influencing host resistance is genetic variation from different geographic location and six such biotypes are known with biotype 4 being virulent (Benett *et al.,* 2004). The recent discovery of two hymenopterans causing galls on *Eucalyptus* and *Erythrina* has caused serious concern. *Leptocyba invasa* causes serious injury to young foliage forming bump-shaped galls on the leaf midribs, petioles and stems of new growth of eucalyptus. Similarly another invasive species *Quadrastichus erythrinae* causes severe galling on *Erythrina stricta* grown throughout the tropics as wind breakers for soil and water conservation, besides being used as a standard for peppers, betel vine and vanilla and also as a favourite fodder crop. Galls are formed in the stem, petiole and leaf lamina resulting in malformation of infested leaves, which fails to attain normal size, with the petioles and tender stems presenting a curly appearance, with knot-like galls, ultimately resulting in cessation of growth. Even the teak crop is not spared, *Asphondylia tectonae* Mani, causing stem and leaf galls. Sandal plantation also provide instances of leaf galling by thrips such as *Crotonothrips gallarum*.

Further, many insect manipulate or alter plant parts tending to increase the availability or quality of resources. In most cases gall formation has a deleterious effect upon the host plants, changing plant architecture, shortening leaf longevity and reducing photosynthesis, seed production, biomass etc. Such deleterious effects become very serious when gall density is high resulting in a shortage of galling sites for the subsequent generations (Yukawa, 2000). Using gall inducing arthropods in the biological control of weeds is another aspect of economic cecidology. Needless to highlight that galls offer economic products such as phenolic compounds like gallic acid and tannins, besides provides species specific chemical signals.

Inquilines in galls

Insects incapable of inducing a gall, but are endowed with adaptations for exploiting shelter and nutrition made available by gall insects are *inquilines.* They are temporary inhabitants not participating in gale formation but subsist by feeding on gall tissues. Thrips galls have generally a single in quiline, but more the one are seen in other galls. In cecidomyiids and cynipids numerous inquilines are known, constituting 6-8% of the total diversity of gallers (Yang *et al.,* 2001). Inquilines range from commensals having little or no impact on the gallers to natural enemies of gallers. For a better appreciation of the biology of inquilines, the need for an understanding of the ratio of inquiline species to natural enemy and inquiline species richness is necessary (Sanver and Hawkins, 2000). In short, inquilines play an important role in community structures.

Conclusion

The evolutionary interactions between a plant and an insect unfolds a sequence of events such as alteration of adult emergence, mating and oviposition times which tend to markedly reduce gene flow between populations associated with different host plants. Gall making insects are excellent model systems for exploitation of speciation processes. Besides efficient resource partitioning and effective utilization of the hosts are very typical of gall insects. The gall forming tendency appears to be a major offshoot in relation to phytophagous habits involving host selection and adaptive strategies resulting in host selection patterns, modifying host tissues to provide for adequate nutrition and reproduction. The insect organized gall systems appear stable and the stability is so apparent that it prompts the idea that gall insects solicit genetic expression from their host cells. As such an interplay of two unrelated organisms as in the insect gall systems, highlights the evolution of a remarkably efficient interacting systems. Plant infestation levels are determined in part by resistance mechanisms, preference and performance, resistance and tolerance, being expressed in sequence. Needless to indicate that a study of the adaptive radiation of galling insects is an interesting aspect of the study of the taxonomic relatedness of host plants utilized by gall insects. The effects of interaction of gall insects, besides injury to cells and tissues, osmotic changes and nutritional alterations, also involve disturbances in their chemophysiological functional capacities. As indicated by Price, "uniting the subjects of behaviour, ecology and evolution, the study of adaptive radiation of gall inducing insects, will no doubt remain a fascinating challenge".

There can be no greater appreciation towards the cause of Cecidology than the recent comprehensive two volume work on the biology, (Raman *et al.*) ecology and evolution of gall inducing arthropods which highlight the structural functional and applied aspects of Cecidology. All the same Mani's work on *Cecidozoa* and *Zoocecidia* and *Ecology of plant Galls* have paved the way for a better appreciation of the viscititudes of galling phenomena, involving ecological integration of the gall inducer and host plants which have kindled the imagination of subsequent workers all over the world.

CHAPTER - VII

INVASIVE INSECTS

Non-native species with potential to threaten `ecosystem' habitats or species, which establish a viable population in an area where it does not normally occur are known as `invasive species'. In short, they increase the probability of a species establishing in habitats that provide food and climate similar to which the species is adapted. They are more likely to be widespread and abundant in their native range with a high capacity for population increase and can function in a wide range of conditions, with a strong dispersal capability. They rapidly synchronize their life cycle with their new environments and have much greater genetic variability (Worner, 2002). Successful invaders are those that can disperse easily, reproduce and develop rapidly, readily find mates or are parthenogenetic, adapt to extremes of weather, new host plants and pest management practices and can survive at low population densities (Baker, 2002). To be described as invasive, a species has to pass through the introductory stage, establish a viable population and cause economic or environmental harm. Knowledge of both phenotypic and genetic variation of species are useful, since those with genetic variability enables it to withstand environmental fluctuation, adapt to wider habitats and make them good invaders (de Castri, 1990). Insect species can often maintain small, viable populations and such species tend to reach outbreak proportions when conditions are favourable for causing economic damage. The rate at which invasive species colonize depends on the capacity of the insect to disperse on its own or through human activities. Insects which are able to function in a wide range of environments are considered good invaders. Besides, uniparental reproduction is a useful feature for an invasive insect, allowing rapid population increase and good invaders have the capacity to rapidly synchronize their life cycle.

As to the characters of invasive species, they are more likely to the associated with human activity invading new areas, and as in aphids, they are widerspread in their native range and are sufficient abundant, with the capacity to invade other areas.

A number of important agriculture pests spread at varying rates, as in *Sogatella furcifera* and *Nilaparvata lugens* which migrate long distances to occupy virtually all suitable habitats under favourable conditions. Both ecological and intrinsic fasters influence the distribution of alien crop pests. Climate, soil, host plants, vector, natural enemies and competitors are the main ecological factors, while genetic adaptability, reproductive strategy, relative humidity, and soil moisture are intrinsic factors. Temperature is the principal factor limiting the distribution of alien species, which also influences physiological processes that are critical to the survival of invading populations, as exemptified by warmth loving species such as *Franklinella, Liriomyza* and *Bemisia* which are unable to complete their life cycles outside protected environments.

The ecological correlations of invasive potential are large original range, commoness, broad diet and short generation time, besides high genetic flexibility, gregariousness and wide climate tolerance-Deliberate introduction of exotic species should consider its ability to spread for several years after release. Of significance in the fact that populations of introduced species often quickly attain outbreak levels on a new host plant in the absence of their natural enemies. In short, large scale ecosystem perturbance of diverse intensites and land transformation provide ideal conditions for invasive species to colonize and multiply. Intrisnic factors common to successful invaders relate to their ability to disperse easily, subsequent to surving ther ecological thresholds. The rate at which invasive pests colonize, depends upon the capacity of the insects to disperse on their own or through human activities. Chronic invading species like the oriental fruit fly *Bactrocera doralis,* are major pests in their endemic source habitats. More than ten invading fruit flies are known and captured in California each year, and outbreaks of many these species have provided considerable ecological information in the more susceptible habitats.

In recent years the invasive gall inducer on Eucalyptus, *Leptocybe invasa* forming bump-shaped galls on the leaf midribs, petioles and stem, has found its way into Tamil Nadu from Australia, via Middle East and Africa, and has become a serious pest of young Eucalyptus plantations. Similarly the *Erythrina* gall wasp *Quadrastichus erythrinae* has invaded India from Mauritius causing malformation and swerl defoliation and death of trees. Erythrina is grown as a wind break for soil around plantations, besides as a standard for black pepper to vanilla. Proliferation of tissues in the attacked region causes gall formation on the entire stem, petiole and leaf lamina with profound enlargement and malformation. Multiple galls with layers of larval chambers are formed in the galled region. Infested leaves shrivel with thick galls on them. Petiole and tender stems enlarge in thickness presenting a curled appearance with knotty or knot-like galls. Records show that 271 wasps emaged from a single infested leaf. As the newly emerged leaves are converted into galls, there is a severe reduction in the number and size of leaves and cessation of growth (Faizal *et al.*, 2006). Similarly, the massive presence of galls in young, encalyptus plantations and nurseries cause serious economic damage. The rapid spread, fast population growth probably as a result of thelytokous reproduction, multivoltinous development and absence of principal natural enemies (Mendal *et al.,* 2004).

The ability to invade or upset native ecosystems, the rapidity of their spread, their population diversity, are important aspects that need consideration. Equally important are the need to recognize new associations formed by phytophagous species, expansion of ecological ranges in the new environment and genetic plasticity. Host switches whereby native insects change their habitat to feed on exotic species are evident in *Iridomymex humilis*, a native of Argentina and elsewhere, is an extraordinary aggressive colonizer destroying most native ants in colonized areas (Elton, 1958). Similarly the vespid *Vespula germanica* continues to show a much broader global expansion.

The invasive species on entry into new habitats dominate subsequently, causing damage to the native fauna. International introduction and establishing effective natural enemies from the native home range of the invasive species is an ideal way of managing invasive species. Successful effort in this direction include *Rodolia cardinalis* to control the cotton cushion scale *Icerya purchtasi* and *Cryptochaetum iceryae* are among the early efforts to promoting classical biological control for invasive insects. (2) The host plant of the psyllid *Leucaena leucocephala* was also introduced into India from Central America and the psyllid *Heteropsylla cubana* has possed a severe threat to the cultivation of *Leucana* all over the tropics. In Karnataka *Leucana* is cultivated in an area of ten thousand hectares and further cultivation was abondoned due to fear of loss of plantation due to psyllid attack. High fecundity rate of 400 eggs and development are completed in ten days. The ladybird beetles *Curinus coeruleus* was introduced from Mexico to Thailand and this predator has since established in various parts of South India.

3. The serpentine leaf miner *Liriomyza trifolii,* an agromyzid entered India accidentally in 1990-91 and has invaded Andhra and Karnataka infesting several host plants including castor. Being highly polyphagous it has affected more than a hundred hosts, and is a serious pest of greens, cucurbits, tomatoes, castor and ornamental plants.

4. The coffee berry borer, *Hypothenemus hampei,* a native of N.E. Africa, was first noted in Gudalur in 1990 and introduced accidentally through coffee brought from Sri Lanka has spread to many coffee growing areas.

5. The spiralling whitefly *Aleurodiscus dispersus* was introduced into India in 1994, so called because the stalked eggs are laid in an irregular spiral around with white waxy material. Adult population build-up on avenue trees and the sticky honey dew produced enables growth of suity moulds.

6. *Bemisia tabaci* `B', the silver leaf whitefly known as one of the world's worst invaders. infesting over 900 host plants and transmitting 11 viruses species, has become invasive through transport of plant products. This was first noticed in 1999 in Kolar district of Karnataka and subsequently invaded all over. It kills hosts by feeding and addition of massive quantities of honey dew which encourage growth of sooty moulds on leaves, inhibiting photosynthesis.

Many more examples of invasive insects are known, though not highly destructive as the ones mentioned above, and Diptera, Hemiptera, Hymenoptera, Lepidoptera and Coleoptera are transported by air/ships, enabling their introduction. Invasive alien ants has been destructive to many ecosystems and the most polymorphic ants are more successful because of feeding flexibilities. The most widerspread ants have been called `tramp species' and this geographic spread in ceased with human activity and are most frequently encountered in urban environments. At the ecosystem level atleast one ant species *Solenopsis wagnetis* has decimated areas when it is introduced. In some cases invasive ants have disrupted ant-plant mutualism. Introduced ant species can serve as excellent bioindicators for assessing the status of an ant community (Schultz and McGlynn, 2000). Introduced ants species can serve as excellent bioindicators for assessing the status of an ant community.

The interception of insects at ports of entry involves a host of orders like Coleoptera, Hemiptera, Lepidoptera, Thysanoptera and Diptera. Among the Lepidoptera are noctuids, pyralids, geometrids and gelactidis; Among the Homptera are mirids, aphids, cicadellids and coccids; curculionids and scolylids among the Coleoptera and agromyzids and tephritids among Diptera. The interrupted insects vary from 500 to 7500. *Ceratitis* and *Dacus* species among Diptera are commonly interrupted, not to mention of *Bemisia tabaci* the world's most serious vector of more than 27 different plant diseases - a consequence of increased trade and human movement.

In order to establish in an area, the need for temperature suitable for reproduction and development, are important. While exploiting intermediate conditions to grow and reproduce, successful establishment occurs only if they can survival the extremes of weather. In other words the climate suitable for colonization is important. Biotic factors which may limit their distribution, are availability of host plants, competition and predation. The role of competition and natural enemies is in preventing establishment of invasive species is important, in view of some ecosystems and communities being more vulnerable to alien invasions.

Alien species have been introduced at all levels into both trophic and mutualistic networks. The impact on native species as competitors, prey species, predators, pollinators, and parasites are not fully known and need more intensive studies. Rarely, it so happens that a parasitoid maintained at high densities on a common pest insect can potentially drive a rare, non-target species to extinction (Stewart *et al.*). Several bee species such as *Apis mellifera*, *Bombus, Megachile* have been deliberately introduced around the world for crop pollination and for honey. Decline in the native bee population occurs where exotic bees are present. To quote Simberloff (2004).

Most key issues in invasive biology fall squarely at the community leval" and to answer how an invader affects a community ``often entails ingenious, detailed research on complicated systems because many inspacts... are subtle, even if devastating."

Biological invasion provides a novel opportunity for the study of life history evolution in novel habitats (Thompson, 1998). In particular, these populations offer the possibility to identify traits of these invading species, that promote colonization and establishment. Dimorphic life history with densely dependent population of macroptera facilatates the build-up of population of the delphacid plant hopper *Prokelista marginata*.

When a species colonizes a new habitat, the biotic and abiotic conditions of the new habitat may differ from the source localities, which could alter selection pressure on the new population and ultimately lead to evolutionary change (Ehler *et al.,* 2004). Invasions could be facilitated by phenotypic plasticity for resistance and competative ability. By adjusting its phenotype to the local environment, a plastic phenotype might maintain fitness across a wide range of environmental condition (Franks *et al.*, 2008).

Chapter - VIII

Ecology of weed infesting insect communities and impact on crops

Agroecosystems as a whole present a picture of dynamic interaction, in terms of both crop plants, and weeds, as well as insect infesting them. The abundance, patterns of distributions of various host plants involving crops as well as alternate weed hosts in space and time are important factors in the population dynamics of insects infesting them. Changes occurring in the crop when it matures and the influence of the host plant on the intraspecific competition in insect population tend to be affected by diverse factors. The increasing tendency for a wide variety of weeds to multiply at a fast rate in and around cultivated fields, competing with the crops for nutrients, causing considerable economic losses every year directly through competition for space, light nutrients and indirectly through acting as reservoirs of several pest species, has come to the recognised as in important factor governing agroecosystems. The influence of crop and weed species on the dispersal of insect pests, rate of colonization, population dynamics, feeding behaviour as well as natural enemies appear essential aspects. Populations of a pest species in an agroecosystem are normally dispersed among different species of host plants and since many colonize weeds and breed successfully, their ability to migrate to the crop and inflict serious damage become considerable. An understanding therefore of the rate of colonization of pest species and their population in crops and weeds has become obligatory for an appreciation of the interactions among crops, weeds and insects. One of the finest examples of such interaction relate to the polyphagous species *Haplothrips ganglbaueri* causing sufficient damage to the inflorescence of paddy, occurring in more than fifty species of graminaceous and cyperaeous weeds in and around cultivated fields. A distinct correlation is evident between the population of thrips on the paddy inflorescence, with those of major weeds *Echinochloa crusgalli, Cyperus iria, C. rotundus, E. colona* to mention the major ones. The population density of weeds and paddy are *Echinochloa* 75-90 plants/sq. m; *Cyperus iria.* 50-70 plants/sq. m; *C. rotundus* 60-75 plant/sq.m. Infestation rates of *H. ganglbaueri* in paddy with around 27.5%, 35.3% in *Echinochloa*, 33.9% in *Cyperus iria* and 25% in *C. rotundus*, facilitating easy migration to the rice crops, causing chaffiness of grains (Ananthakishanan & Thangavelu, 1976).

Therefore, field margin and crop edges are highly supportive also of beneficial insects, their relative abundance varying with the mix of plants at the margin. Marginal habitats therefore, provide the stability needed for species and the landscape matrix of field margins are important for effective field dispersal and conservation success. (Dennis & Fry, 1992). Crops edges adjacent to hedges and broad grassy strips bear a rich fauna of beneficial insects.

Weeds may influence insects in diverse ways. (a) by acting as alternate hosts of pest (b) by affecting the biology of pests (c) by affecting pest behaviour and influencing pest morality through harbouring natural enemies (Way & Cammell, 1981). Host plant switching is normally encountered when the crops/weeds dry up, with the exact stage of the host plant unavailable for the insect to feed, besides the tendency to disperse to more nutritious host tissues for feeding and breeding. For example, assessment of the population trends of mirid long *Cyrtopeltis tenus* on tomato, ashgourd, besides the weeds *Cleome viscosa* and *Gynandropis pentaphylla* shows a definite pattern of host plant switching depending upon the availability of the host plants. With the crop being removed from the fields in summer, the bugs disperse to the weed the *Cleome viscosa*. Another example pertains to the weed bug *Spilostethus pandurus* normally feeding on *Gosypium herbaceum* and *Sorghum vulgare,* besides *Sesasum indicum*. Oil seed crops like *Arachis hypogea* and *Sesamum indicum* are infested by *Caliothrips indicus* which builds up populations in the annual weed *Achyranthes aspera*. *Retithrips syriacus* and *Rhipiphorothrips cruentatus* occur on the crops and migrate to the alternate hosts *Croton sparsiflorus* and *Cassia auriculata*. In short, the abundance, patterns of distribution of various host plants involving crops as well as alternate weeds is an important factor in the population dynamics of these species. Ploughing the field at different seasons of the year affects the species composition and abundance of weed complexes and the species inhabitating thems. The role of weeds as alternate hosts tend to have an impact on the biology of the pest species, not mention of their role as reservoirs of pest species, playing a major role in their population build-up during the off season. The diverse management practices, intensive production techniques and indiscriminate use of pesticides have created and ecological pressure on the sustainance of insect pests.

Sap-feeding insect on some weeds

Aphis craccivora	*Cyperus iria*
	Echinochloa colona
	Amaranthus viridus
Aphis gossypii	*Chloris barbata*
	Cyperus iria
	Acalypha indica
	Euphorbia hirta
	Leucas aspera
	Solanum xanthocarpum
	Tridax procumbens
Aphis nerii	*Acalypha indica*
	Datura sp.
	Euphorbia hirta
	Gomphrena decumbens
Caliothrips indicus	*Cyperus iria*
	Cyperus esculentus
	Dactyloctenium aegptium
	Echinochloa colona
	Achyranthes aspera
	Amaranthus viridis
	Amaranthus spinosus
	Cleoma viscosa
	Corchorus aestivus
	Croton sparsiflorus
	Justicia simplex
	Leucas aspera
	Physalis minima

	Solanum xanthocarpum
	Tridax procumbens
Retithrips syriacus	*Chloris barbata*
	Cyperus esculentus
	Echinochloa colona
Haplothrips ganglbaueri	*Chloris barbata*
	Cynodon dactylon
	Cyperus iria
	Echinochloa colona
	Acalypha indica
	Ricinus communis
	Croton sparsiflorus
	Tridax procumbens
Haplothrips ganglbaueri	*Chloris barbata*
	Cynodon dactylon
	Cyperus iria
	Cyperus esculentus
	Dactylotenium aegyptiacum
	Echinochloa colona
	Abutilon indicum
	Achyranthes aspera
	Amaranthus spinosus
	Corochrus aestivans
	Cleome viscosa
	Croton sparsiflorus
	Gomphrena decumbens
	Justicia simplex
	Leucas aspera
	Parthenium hysterophorus
	Tridax procumbens
	Vernonia cinerea

	Parthenium hysterophorus
	Physalis minima
	Ricinus communis
	Solanum xanthocarpum
	Tephrosia purpurea
	Vernonia cinerea
Cyrtopeltis tenuis	*Amaranthus gangeticus*
	Cleome viscosa
	Croton sparsiflorus
	Hibiscus sabdariffa
	Physalis minima
	Ricinus communis
	Solanum xanthocarpum
	Tephrosia purpurea
Nezara viridula	*Cynodon dactylon*
	Cyperus iria
	Abutilon indicum
	Amaranthus viridus
	Amaranthus spinosa
	Cleoma viscosa
	Croton sparsiflorus
	Euphorbia hirta
	Gomphrena decumbens
	Hibiscus sabdariffa
	Leucas aspera
	Physalis minima
	Solanum xanthocarpum
	Vernonia cineria
Franklinella schultzei	*Abutilon indicum*
	Amaranthus viridis
	Cleome viscosa

	Croton sparsiflorus
	Justicia simplex
	Leucas aspera
	Parthenium hysterophorus
	Ricinus communis
	Tridax procumbens
	Vernonia cinera
Amrasca bigwttula	*Abutilon indicum*
	Amaranthus viridis
	Boerhaavia diffusa
	Cleoma viscosa
	Hibiscus sabdariffa
	Leucas aspera
	Physalis mima
	Cynodon dactylon
Bemisia tabaci	*Abutilon indicum*
	Acalypha indica
	Amaranthus viridis
	Amaranthus gangeticus
	Boerhaavia diffusa
	Cleome viscosa
	Croton sparsiflorus
	Convolvulus arvenis
	Euphorbia hirta
	Hibiscus sabdariffa
	Leucas aspera

The abundance, spatial distribution, diversity and sizes of host plants tend to affect the ability of the insect to reproduce and spread. While it is common that most insect pests reproduce best on nutritious crop hosts, their population in a habitat is determined by their capacity to use mixed hosts. The presence of weeds in the habitats being more predictable, the weeds have a major role in integrated pest management. Spatial variation in weed density being considerable, it is common to see dense patches of particular weeds in some areas. As such the spatial distribution pattern decides the competetiveness of the weeds with the crop species. To give a few instances, it has been seen that 26 species of weeds were found associated with the groundnut crops; 31 species of weeds in the cotton ecossytem, 21 species with the tomato crops. In short, 11 monocot and 39 dicot weeds were found to be the associated with the crops cotton, tomato, groundnut and brinjal (Shanti and Jahir Hussain, 2005). The presence of diverse vegetation within or around the fields tend to be essential resources for predators or parasites.

Fig. 8.1 (A-E). Pests and weed complex in Groundnut ecosystem (Courtesy, Jahir and Shanti)

Fig. 8.2. *Aphis craccivora* on groundnut.

Fig. 8.3. Weeds infested by *Spilostelthus pandurus*
(Courtesy, Jakir and Shanti)

Host Plant Switching

Host plant switching is normally encountered when (i) crop/weed dries up (ii) the exact stage of the host plant is not available for the insect to feed on (iii) if there is a nutritional inadequacy so that insect tends to disperse to plants of more nutritive value for feeding and breeding. A definite pattern of host plant switching in evident in the population trends of the mirid long *Cyrtopeltis tenuis* in tomato and ash gourd (*Luffa cylindrica*), as well as on their weed host *Cleome viscosa* and *Gynandropsis pentaphylla*.

The milkweed bug, *Spilostethus pandurus* besides feeding on the crops *Gossypium hirsutum* and *Sorghum vulgare* is recorded as a pest of gingelly (*Sesamum indicum*) (Thangavelu, 1979). The infestation of *S. indicum* occurs just prior to flowering and under severe infestation, the entire leaves of the plant form curled rolls and wither. Early attack by the nymphs on young gingelly plant results in abnormally small leaves with stunted growth and few capsules of smaller contour with severe malformations and seeds poorly developed inside the pods. Under acute infestation the pods dry and wither prematurely. Population studies of *S. pandurus* showed the population to build up steadily just before the harvest of the crop. After harvest the adults fly back to their principal hsot, *Calotropis gigantea*. Similarly, the small milkweed bug, *Spilostethus hospes* was observed causing damage to the seeds of *Solanum melongena, S. nigrum, Helianthus annus* among the crop plants, besides feeding on the milkweed *Calotropis gigantea* and the weed *Vernonia cinerea*. Regular population monitoring indicated the weed hosts to serve as reservoirs to sustain the population until the favourable crop season (Ananthakrishnan *et al.*, 1963).

Another good example pertanis to the degree of infestration of *Chirothrips mexicanus* a seed feeder in young earheads of *Pennisetum typhoides* and the weed *Chloris barbata* in pennisetum fields. The incidence of this species on the weed indicated a range of 25-30 individuals/branch for the four-branched spike, each with 45-50 spikelets, with a preponderance of males, suggest arrhenotokous mode of reproduction, alongwith the sexual one. Of the 176-200 growing florets of *C. barbata*. 57-75 individuals of *C. mexicanus* occured in various stages of development (Ananthakrishnan, 1984) of considerable intact is the cardamom thrips *Sciothrips cardamomi*, whose population ranges from 550 thrips/100 panicles to 800/100 panicles, their being a correlation between thrips population on cardamom and the alternate hosts like *Amomum, Colocasia, Hedychium*.

The study of the influence of crops and weed species on the dispersion of insect pests, rate of colonization and their population dynamics, feeding behaviour, food utilization and influence on natural enemy attraction are essential elements in habitat management.

Weeds are so prevalent that they are an important components of an ecosystem. The fact that many pest species colonize on weeds and breed successfully reveals that they can migrate to the crop and inflict serious damage of any time. This is more so when the growing period of the crop plants synchronize with that of the weeds. Instances where there are definite host plant succession between crops and weeds are on record. Thus a knowledge of the nature of colonisation of insect pests and their population build up on crop and weed host become essential to understand the interactions lectures insect crops and weeds.

Weeds provide a large, probably an overwhelming, proportion of food for the desert locust, *Schistocerca gregaria* and thus this natrual vegetation is effectively exploited by the continually nomadic populations of this pest (Rainey, 1981). In view of the preference of *S. gregaria* for breeding in arid areas that are dry and therefore unsuitable for the growth of crop plants, wild weeds hosts occurring in abundance in these locations assume an important role in the ecology of the desert locsut. Although in general, grasshoppers like *Oxya nitidula,* prefer graminaceous and cyperacous hosts, conforming tot eh graminivorous type of acridids (Uvarov, 1977), the developmental stages show feeding preferences in terms of the first and second instar larvae feeding on weed hosts and the third, fourth, and fifth instar larvae prefering to feed on crop plants. On the other hand, the cotton grasshopper *Cyrtacanthacris ranacea* feeds exclusively on malvaceous hosts including crops like *Gossypium hirsutum* and *Abelmoschus esculentus* as well as on weed like *Abutilon indicum* and *Sida rhomboidea*, displaying a regular movement between the crop plants and weed hosts depending upon their seasonal availability (Ananthakrishnan *et al.,* 1985). Host preference patterns in terms of the rate of development, quantitative food, intake, and growth rate of species such as *C. ranacea* indicate a greater preference to *G. hirsutum;* this host has a greater influence ont he pest in terms of very high fecundity rate. Broadly, food consumption rate is greater on hosts with low water-nitrogen index, a factor that determines post-ingestive growth performance of the insect. Interestingly, *Truxalis indica* feeds exclusively on monocotyledonous plants, more particularly on grasses, *Orthacris maindroni* only on dicotyledonous plants, and *Atractomorpha crenulata* on monocotyledonous as well as dicotyledonous plants (Ananthakrishnan *et al.,* 1985). Significant variations in the quantitative food intake on diverse plants during post shown. Thrips-weed-crops interactions have provided adequate information to be able to appreciate the need for an understanding of such interactions.

The nature of weed-crop interactions with respect to the seasonal fluctuations of insects have indicated the role of the weed *Chloris barbata,* abundant in *Pennisetum typhoides* fields, as an alternate host of *Chirothrips mexicanus.* The inflorescence of *Echinochloa crusgalli* also harbours numerous adults and larvae of *Haplothrips ganglbaueri,* the weed acting as an important alternate host for the thrips in paddy fields. *Caliothrips indicus* an important thrips pest of groundnut, *Arachis hypogaea* is also found to colonize in large numbers on the weed host *Achyranthes aspera* (Daniel, 1984), almost throughout the year, although its infestation on *A. hypogaea* is seasonal. In Tamil Nadu, *A. hypogaea* is grown twice a year, as irrigated and rainfed crop, infestation of thrips was evident on 15-20 days old seedlings. The annual weed, *A. aspera* usually appears by August in *A. hypogaea* fields and acts as a reservoir fo this pest. The density of this weed declines from March anad with the onset of the crop in June, *C. indicus* migrates from *A. aspera*. The peak population of the pest, both on the crop as well as on the weed was evident during September. With the increase in weed density there was a rapid migration of *C. indicus* from the crop to the weed. Similarly, *Zaniothrips ricini,* another thrips pest of castor, *Ricinus communis* has also been noticed to utilize the weed hosts, *Datura stramonium* and *Calotropis gigantea* (Daniel, 1983). On castor, *Z. ricini* appeared during October and by January the peak population was evident with a decline thereafter. This decline of *Z. ricini* population on castor was immediately followed by the appearance of thrips on *C. gigantea, D. stramonium.* In this case both the weeds acted as alternate hosts only for about two months.

The spatial and seasonal distribution patterns of thrips on *R. communis* appear to be of interest. On the castor crop, *Toxothrips ricinus* a highly host-specific species was restricted to the bracts covering the tender buds,w ith stray cases of the larvae occurring on the leaves of the first node. *Scirtothrips dorsalis,* a polyphagous species was abundant on both sides of the tender leaves of the first and second nodes, rarely occurring on the third nodal leaves and inflorescence. The common polyphagous species, *Retithrips syriacus* mostly occurs on the third leaves downwards. In ddition a meagre population of *Rhipiphorothrips cruentatus* was also present from the third leaves downwards. In addition a meagre population of *Rhipiphorothrips cruentatus* was also present from the third to seventh nodes along with *R. syriacus.* A similar low trend of population was observed in *Astrothrips tumiceps. R. ricini* infests the lower surface of the leaves from the fourth to the seventh nodes, while adults and larvae of *Ayyaria chaetophora* exist from the fourth node onwards. From the third to the seventh nodes, a mixed population of *Astrothrips tumiceps, Ayyaria chaetophora, Retithrips syriacus,*

Rhipiphorathrips cruentatus, and *Zaniothrips ricini* was observed (Daniel 1984; Ananthakrishnan, 1984).

Similarly, on the weed host *Achyranthes aspera, C. indicus* appeared to select a definite site such as the spikes for feeding, resting and reproduction. Another species *Frankliniella schultzei,* the pollen-feeding anthophilous species occurred along the adaxial sides of the first nodal leaves of the plants bearing inflorescence. This species was completely absent during the non-flowering season of the plant. The second nodal leaves were generally occupied by *A. chaetophora* and *C. indicus. C. indicus* prefers the second nodal leaves for egg-laying, moving to other nodal leaves for feeding. When the second nodal leaves become third node due to the production of a new pair or leaves, the gravid females migrae to the fresh second nodal leaves for egg laying.

Effect of Wild Plants on the Biology of Insect Pests

Among the various groups of insects, aphids are very much renowned for their complicated life cycles. About 10% of species alternate between a primary host, on which the sexuals give rise to overwintering eggs and their resultant spring generations, and secondary hosts, on which parthenogenetic generations spend summer and eventually give rise to autumn migrants, which return to the primary host (Estop, 1981). Most alternation occurs only in some subfamilies and may be achieved in different ways in different subfamilies. For instance, alate males of alternating Aphidinae, fly from secondary to primary hosts, while in other groups sexuparae fly from the secondary hosts to produce apterous males on the primary host. While most alternating aphids return annually to the primary host, almost regardless of conditions, some individuals may remain on the secondary host in the autumn reproducing parthenogenetically.

The nature and extent of survival, growth and reproduction of insects on crop and weed hosts ensures not only the suitability of the crop for the colonizing species of insects, but also reveals the ability of different weed species to harbour different species of insects. The growth rate and fecundity of teh lygaeid, *Oxycarenus hyalinipennis* and the mirid, *Cyrtopeltis tenuis* showed significant differences when colonized on crop and weed hosts (Ananthakrishnan *et al.,* 1982). Apart from cotton, *O. hyalinipennis* feeds on a number of alternate malvaceous host plant seeds viz., *Abelmoschus esculentus, Abutilon* sp. *Sida* sp. etc. The growth rate and fecundity were highest when reared on cotton seeds, than on other host plants. Similarly, the survival, post-embryonic development and reproductive efficiency of the tomato mirid, *C. tenuis* on crops and weed plants as effectively as the crops. Eventhough from the nutritional point of view the weeds were less preferred as compared to crops plants, *C. tenuis* was able to feed and reproduce thereby utilizing the feral host plant as a reservoir.

Weeds Harbouring Natural Enemies of Pests

Many instances can be cited with regard to natural enemies with free living stages that feed on the pollen and nectar of wild plants and with predatory/parasitic stages tha feed on prey or hosts on wild plants. Flowers of many weed plants provide sumptuous food for natrual enemies, for example, adults of some Syrphidae and parasitic Hymenoptera. The giant ragweed, *Ambrosia trifida,* is recognized as supporting a host of the first generation of the parasitie, *Lydella grisescens* which then caused notably increased parasitism of European corn borer, *Ostrinia nubilalis* (Hasio & Holdaway, 1966). *A. trifida* also supports an important host of parasites of boll weevil, *Anthonomus grandis,* and orient fruit moth, *Cydia molesta* (Zandsra & Motorka, 1978). The syrphids, *Melanostoma* spp. laid more eggs on brassica plants in weeds than on weed-free brassicas (Smith, 1969).

Although several researchers have demonstrated that certain weeds serve as reservoirs of alternate hosts and prey for natural enemies (van Emden, 1965; Hamenway & Whitcomb, 1967; Flaherty, 1969; Perrin 1975), only a limited number of weeds appear to be tolerated in fields of cultivated crop plants, since the weeds can hamper the economic threshold level and compete with the crop. Hence the trend envisaged now is to maintain certain specific weed associations in crop areas to provide supplementary food for entomophagous insects (van den Bosch & Telford, 1964). However, the objective of maintaining a desired weed composition at a limited density has been seldom achieved. Doutt and Nakata (1973) advised the use of weeds in vine yards of California to manage the grape leaf hopper egg parasitoid *Anagrus epos.*

Many of the predators can be increased through seasonal manipulation of selected weeds and effectively used as natural enemies of important crop pests. Early stages of crop growth, the presence of specific weeds may act asa reinforcement for natural enemies of crop pests and thus sustain the population of predators prior to the onset of the pest.

Chapter - IX

Insect Guilds : Structure and Function

The concept of insect guilds was introduced by Root (1967) for `groups of species that exploit the same class of environmental resources in a similar way'. In today's parlance guilds are defined as areas of interspecific competition and resource partitioning. It is well known that complex and numerous interactions among species are characteristic of real communities. Such "resource oriented organizational units within communities that transcend taxonomic boundaries" are referred to as *guilds* (Hawkins and McMohan, 1989). Evans and Murdoch (1986) stated that "the feeding action of organisms impose a pattern on the community which overrides taxonomic compositions". Terborgh and Robinson (1980) opined that guilds will become the standard currency of ecologists in their effort to understand community relationship". In short, guilds are groups of functionally similar species, the fortunes of different species within the guild varying individually in response to factors other than resource availability (Hawkins and McMohan, 1989). The guild concept tends to unity community and ecosystem approaches to ecology, different guilds exploiting different types of resources (Root,) has suggested that "guilds represent major ecological categories that have been moulded by adaptation to the same class of resources".

Gould (1973) divided folivores of collards with three district guilds based on the modes of attacking plants - *pit feeders, strip feeders* and *sap feeders.* Following this several other guilds have been suggested such as chewers, miners, parasites, wood borers, stem borers, nectarivores, omnivores, fungivores, epiphytic grazers and dung feeders (Hawkins and McMohan, 1989). Guild composition varied considerably between sites both in relation to numbers and in the biomass of individuals. Guild structure varied significantly between narrow leaved and broad leaved plants, the former having more chewers and sap suckers. As plant architectural complexity declines, so too both the diversity of guilds. Plants with greater variety of resource type such as feeding, resting overwintering and hiding places support more species of herbivorous insect (Lawton, 1983).

Successional shift in guild composition of more narrow environments such as of dung pats in known. Dung specialists (coprophages) when the first beetle to colonize droppings, with subsequent appearance of more generalists (Hanski and Korkela, 1977). Such dung pat beetles were classified on top specialist, bottom specialist, periphery and central specialists. These patterns of dispension are explicable in term of the rarity of dung beetles in relation to their chances of mating.

Stem boring guild species of herbivorous plants are well known, viz. specific plant exploitation patterns of a guilds of 48 stem boring insects. The patterns of dominance, diversity, species area and dispensive for a guild of herbivorous boring insects and their plant resources, diversity pattern based on competition have been examined (Rathke, 1976). The high overlap in resource exploitation among guild members reflect the selection for competitive avoidance. The stem boring guild comprising cerambycids, curculionids, cecidomicids, Tephritids, pyralids is well known. A major advantage of a guild approach is the importance of competitions to community structure. As such guilds are considered as "arenas of intense interspecfic competition with strong interaction within guilds, but weak interactions between members of different guilds, enabling a better understanding of all aspects of competitions (Hawkins and McMohan, 1989). The formers of different species within guilds vary individually in response to faster other than resource availability.

Guild composition tends to vary considerably between sites, both in relative numbers and biomass of individuals. Guild structure varied significantly and the patterns of dispersion are explicable in terms of the variety of dung beetles in relation to their mating choices (Holter, 1982). Competition was observed between endocoprids living and feeding within the dung heaps and species that roll dung both. A good example relates to guild comprising 13 species *Aphodius* that coexists in cattle dung pats. The entire *Aphodius* material comprised 39000 beetles. The contagious distribution of beetles on evenly spaced identifcal pats, the positive association between species and habitat specialization within pats, may have evolved as adaptations facilitating intraspecific interactions in populations (Holter, 1982).

Patterns of guild structure are very typical of canopies, involving xylophagus, herbivores, saprophagus, predators and fungivores. Variations in the proportional abundance of different guilds are known, which tend to vary with a high degree of tree variations.

Tourist guilds are known, which have no lasting or intimate association with the plant, but which form part of a tropic web in the arboreal community. Dipteran guilds are well known and involve the fungal guild (Mycetophilidae and Sciaridae), phytophagous guild (Cecidomyiidae and Tephritidae), Predatory guild (Euphiididae, Dolichopochidae and Ceratoprogonidae) and parasitic guild (Tachinidae and Pipunculidae). The ephemeral nature and non-specificity of interactions make them tourists.

A major advantage of a guild approach is the importance of competition to community structure. As such guilds are considered as "arenas of intense interspecific competition with strong interaction within guilds, enabling a better understanding of all aspects of competition. Niche related to a wide variety of ecological aspects including size-ratio difference within guilds. If each guild exploits different types of resources, guild composition of communities should reflect the availability of resources (Hawkins and McMohan, 1989). The fortunes of different species within a guild vary individually in response to factors other than resource availability. In short, guilds are of utilitarian value in ecological studies, since they share the same resources, occur in the same or overlapping microhabitats and are taxonomically related. What is desirable is a functional classificatory approach to the guild concept, using a broader prospective involving a direct comparison of the components of similar communities in different places at different times (Hawkins & McMohan, 1989).

Parasitoid guilds

Insect populations are regulated through interaction between insects and their natural enemies and habitat stability of ecosystem such as cropping system tends to initiate the establishment of natural enemies. This is true of exotic pests where they are mine successful such as noticeable in forests. Host density dependence, host susceptibility and variable parasitoid sex-ratios are essential factors. Parasites have an innate limited ability to find grey upto a certain density and mutual interference results in high parasitoid densities, with a significant stabilizing effect as low populations densities of parasitoids and hosts (Dent, 1993). Community ecology therefore, refers to the structure and diversity of parasitoid communities, with parasitoid complexes, their host range and variability in the number of parasitoid species. Alterations in natural ecosystem through human interference results in patterns and processes in parasitoid communities. The way in which a resource or a particular food is utilized by *parasitoid guilds* or groups of species sharing the same niches is important. For example, egg parasitoids, larval ecto- or endoparasitoids pupal, prepupal and adult parasitoids form distinct guilds. As such parasitoid communities are to be considered as a set of distinct guilds with increased competitive interactions as evident from their richness in the different guilds. With the evolution of plant communities, their hetrogenecity also changes and necessarily insect hosts of parasitoids also change. Mirids, delphacids, cicadellids support poor parasitic communities as compared to Lepidoptera. Parasitoids have played a significant role in shaping the determination of species richenss. Three types of parasitoid webs are known viz. connectance, semi-quantitative and quantitative. An interesting example, relates to *Hyblaea peura* the teak defoliator with 15 species of parasitoids, seven from larvae and rest from pupae. Eulophids, *Elasmus* sp., *Sympiensis* and a tachinid *Palexorista,* besides maximum parasitisation by the richness of *Eriborus gardneri. Stictopisthus* is well known parasitoid. By aggregating in certain patches of high host density, the parasitic leave other patches as partial refuges parasitisation (Nair *et al.,* 1995).

Foods webs are classified as:-

Source webs beginning with a primary producer or herbivore and include all higher trophic levels.

Sink webs are centred on one particular predator and include all members of lower trophic webs.

Community web includes all species in any defined community.

Food webs not only reveal trophic interactions between different parasitoid species, but also involves competitive interaction between them. Parasitoid guilds are ecological units, not a taxonomic one and guilds are often considered to be the building blocks of ecological communities and plays a central role in community ecology. The parasitoid guild associated with the cecidomyiid gall midge *Rhopalomyia* are Torymidae, Pteromalidae, Tetrastichidae, Eulophidae, Eupelmidae and Platygastridae. A natural guild consists of native, covered species which exploits the host species in its native home. When one or more exotic species involves the natural guild, it becomes restructured. The control of population depend on the number of species in a guild.

Chapter - X

Insect and Climate Change

The earth's climate has always been subject to natural variation with shifts between cold and warm, but the current warming trend ecologists worried for reasons more then one, this is the first instance when man appears to accelerate the change enabling warming to take place quickly, so that species will not have time to adapt and adjust and avoid extinction. Since different species adapt to climate change in different ways, the natural cycles of interdependent species tend to fall out of synchrony. In short, things that normally happen in geological time are happening during the span of human life. Habitats are changing along with the timing of annual processes the flowering, leaf coverage, migration and birth. Species that depend on each other like planet and insects that pollinate them, do not always adapt to change at the same time. In short, environmental changes will alter both the timing and rates of interaction with food plants, competitors and natural enemies. Uncoupling of certain associations and establishment of new ones are in the offing, as species responds to changing climate conditions. Loss of genetic variability in the newly established host ranges tend to influence the ability of insect population to adapt to changing climatic conditions involving new food plants, competitors and natural enemies. In short, climate change will affect incidence and severity of limiting factors as well as the favourableness of growing conditions.

The key factors in changing earth's climate are the greenhouse gases released by industry, agriculture, automobiles and coal fired electrical generation plants. We are now the geological agents capable of affecting the process that determines climate. The concentration of CO_2 in the atmosphere helps to determine the earth's surface temperature and CO_2 and temperature have risen sharply since the 1950s, notably though forest clearing and fossil fuel burning which have pushed up atmospheric CO_2 by nearly 100 parts/million and the 20th century has seen the greatest warming in a thousand years. In the context of climate change, consideration of the atmospheric CO_2 is important, besides annual cycles of temperature, moisture, light and wind. The key variables are maximum and minimum temperature, rainfall and evaporation, day length and their seasonal patterns which have their impact on insects and such studies on climatic change on insect population dynamics important. Porter *et al.* (1991) list of effects of temperature on insects which limit their range, over wintering, population growth rate, number of generations per annum, length of growing season, crop-pest synchronization, intraspecific interaction, dispersal, migration and availability of host plants and refugia.

The entire chemical environment of the biosphere is being transformed by man through the development of new technologies. As temperatures rise, several cues are altered and life tends to shift in space and time. Habitats are changing along with the timing of annual processes as indicated earlier. Climate warming of about 1-2 c would increase this tendency in relation to larval emergence and bud burst. As such a need for the study of combined effects of several factors of multi impact studies related to CO_2, temperature, pollutants such as ozone become important. Such studies need to have priority if we are to establish how different forms of plant feeding insects respond to such changes, notably the impact of global temperature rise in relation to abundance and distribution of plant feeding insects. A study of 35 non migratory butterflies species in Europe found in recent decades, that about two-thirds have expanded their range by 20-150 miles and many plants flowers a week earlier then they did 50 years ago. So as temperature rise, seasonal cues are altered and life shifts in space and time. Increase in C:N ratio tends to have a detrimental effect on polyphagous insects leading to increase in carbon based allelochemicals. Further, the negative response of insects to elevated CO_2 is that they accumulate more carbon, with an increase of C:N ratio, so that insects are faced with nutritionally deficient host plants resulting in a N-dilution effect.

Climate and biotic interactions

Seventy years ago Sir Boris Urarov, the renowned Entomologist and locust specialist, in his classic paper on 'Insects and Climate` discussed the problem of predicting the response of insect population to environmental changes indicating that "The climate of every area on the surface of earth has undergone a long series of great changes throughout geological history. Patterns of glacial advances and retreats, changes in sea levels, changes in rainfall patterns all have had their effect on earlier civilization, but the difference in the changing climatic patterns in recent years have been augmented at an unusually fast rate, thanks to the anthropogenic environmental changes, mostly degradation or destruction of the world's natural habitats, as well as the combined efforts of habitat loss, fragmentation, pollution and climate changes which have posed a serious threat to the world's biodiversity". A good example of total extinction of a species through landscape fragmentations the sudden disappearance of the locust *Melanoplus spretus*. Since the geographic distribution of species is a consequence naturally elated to local abundance in the favorable parts of the habitats. The climate of a given area being partly a function of the organisms in the area, the capacity to tolerate various climatic conditions during unfavorable seasons is a characteristic of several species of insects. While individual species respond independently to climate, the sum total of individual responses makes up the response of communities. The occurrence of short periods of extreme climate is often lethal to species, but their abilities exploit increasing vaguries of climate as well as the potential of phytophagous insects to expand their host range are aspects which need increased scrutiny. A species may adapt completely to one extreme environment even at the expense of near lethality in the other it may also be polymorphic containing a mixture of polymorphs. Some of which are adapted to each environment. Besides rapid changes in climate could disrupt forests and other ecosystems, reducing their to draw carbon dioxide from the atmosphere. Soil microbes continuously decompose the carbon in CO_2, and methane which tends to enhance warming and seriously disrupt biotic interactions, taking into consideration the possibility of doubting of atmospheric CO_2, in another 50 years.

The exploitation of natural resources at a fast rate involving initiation of large scale developmental schemes and multipurpose projects and different types of land use and new farming systems and other agroforestry practices have considerably altered or disrupted the environment of local insect faunal complexes with consequent change in their composition. There can be no better examples than the influence of the Rajasthan canal on the increased incidence of crop pest and insects vectors, as against the decreased incidence of the dreaded red hairy caterpillar, *Amsacta albistriga* in Tamil Nadu, with the advent of the Parambikulam - Aliyar dam. Though fossil records show long term stability of insects, it should not be surmised that they are invulnerable to environmental changes. Tropical deforestation could bring about essentially different ecosystems through its impact on local climate. As has been clearly indicated by Sprugel, human influence and small climatic variation can cause very substantial changes in vegetation, and the biodiversity for any given landscape will vary substantially over any significant period'. The variety of genes, species and ecosystems evident in present day forest ecosystems result from a combination of cyclical, ecological and climatic processes and past human actions. While the effects of deforestation and decreased forest floor insect diversity are well known, the introduction of increased monoculture has also seriously interfered with insect faunal complexes, not to mention of land use changes and resultant increased changes in species composition, diversity, numbers, and biomass of insects. Environmental fluctuations occurring over long periods of time are likely to result in changes of dominance among the species, organisms themselves often causing environmental changes (autogenic), as against the changes caused primarily by external factors (allergenic). Change in environmental conditions associated with long term climatic fluctuations can produce allergenic successional changes in terrestrial vegetation and associated insects. Whether climatic changes are gradual or sudden, most species have the tendency to become pests, notably the scarcer ones and for a long time the existence of a link between outbreaks of insects pests and atmospheric pollution was suspected. To quote Robert Metcalf, ``it seems inevitable that by the end of the 21st century, all of the world's most notorious insects pests will have a cosmopolitan distribution wherever suitable climate and hosts exists''.

Environment of crops and risks of climate change

Effects of climate change in an agricultural environment for crop pests (essentially include direct effects of carbon dioxide (CO_2), nitrous oxide (NO_2), methane, chlorofluorocarbons commonly from many sources including wetlands, rice fields, landfills which together increase the warming effect of CO_2. Changes in the quality of plant species are brought about as a result of increase of CO_2 and temperature induced stress, which tend to result in altered growth, development and reproduction in insects. CO_2 now contributes about 350 ppm and such a dramatic increase has a direct impact on global warming, naturally affecting cropping systems and insect behaviour, as has been very evident in horticultural crops, Stomatal closure and reduced transpiration would tend to result in higher leaf temperature, reduction in RH, which are not favourable to development in some insects. Increase n the carbon dioxide to nitrogen ratio of plant foliage is most likely to have a detrimental effect on insect herbivores, since it is possible that this could be due to increased levels of allelochemical, besides the nutritional quality of the hosts. The expression of constitutive host plant resistance varies with environmental changes which may also be known to be affected by climate, so that environmental factors influence the magnitude of expression of host plant resistance.

Carbon-dioxide, one of the greenhouse gases emitted is a major factor attributed to changing patterns of plant distribution and homeostasis. Increase in CO_2 favours increase in the rate of plant growth and photosynthesis and decrease in nitrogen fixation by the plants. The cause of concern is with C_3 plants (wheat, rice, soya etc) which re-oxidize CO_2 during photorespiration unlike the C_4 plants (maize, sorghum, sugarcane, millets) which are less affected. Thus, the C_3 plants have to respond to the stress positively by suppressing the rate of photorespiration. A shift in the demographic distribution is likely to occur with the persistence of low yielding crops like C_3 plants over high yielding C_4 plants. Carbon-dioxide increase may not only influence the plant selection pressure, but also alters plant-insect interactions. By the year 2050, soybean (*Glycine max*) the world's most widely grown seen legume is presumed to grow in a stressful atmosphere with a 50% higher CO_2 and could affect plant-insect interactions. Soybeans respond to insect attack by producing defensive compounds that inhibit digestive enzymes (proteinases) which are under the influence of Jasmonic acid in the gut of insects, reducing their performance and crop damage. The impact of elevated CO_2 directly on the abundance of phytophagous insects has been studied on some insect species and *Pectinophora gossypiella*, *Empoasca* and *Trialeurodes vaporiorum* were most affected.

During the process of uptake of CO_2 from the stomatal openings in the leaf, the leaf releases water through a process termed as transpiration. Doubling of ambient CO_2 causes 40% decrease in the population of both C_3 and C_4 plants reducing transpiration by 23-46%. Increased CO_2 levels may encourage foliar growth and canopy leading to moisture accumulation and enhanching pathogen growth. For e.g. soil microbes and symbiotic organisms will be more vulnerable leading to loss of soil fertility.

Increasing warming and changing rainfall patterns tend to lead to loss of soil fertility, resulting in a decline in agricultural productivity, not to mention of increased instances of water borne diseases due to increase in vector dynamism. A notable example in the extreme vulnerability to the impact of climate change in Orissa. In view of development and climate change being closely interlinked, the need for a better understanding of water resources, agriculture, forestry, human health and industry becomes obligatory. Green house gas emission becomes important, involving CO_2 emission, emission of methane, NO_2, hydrofluoro carbons and sulfur hexafluoride, besides their impact on climate change. In particular, growth and yield responses of important crops in terms of productivity and loss due to increase in pest load occur. In short, the importance of crop-pest-weather management interactions becomes important. Crop-soil-water-nitrogen-pest-climate interactions form an efficient tool to understand the impact of climate change in agricultural productivity.

Many sensitive ecosystems rich in biodiversity are very vulnerable to climatic change. Climate change could increase soil degradation, rate of soil erosion hampering food production. Moisture availability in term of climate change in important. Mismanagement and depletion of natural resources, altered resource available due to regional and global climate and environmental change tend to affect the migration pattern of insects. Increase in carbon and nitrogen ratio of plant foliage is likely have a detrimental effect on insects leading to an increase in carbon based allelochemicals.

The impact of elevated CO_2 has been evident in a decline in foliar water content resulting again in an increase of carbon compounds. Decline in the nutrine quality of plants tend to result in poor performance of insects with the tendency to become less abundant as atmospheric CO_2 levels increase, with increased possibilities of species extinction.

Homoptera and Climate change

Aphid biology is influenced by temperature which initiates development, fecundity, movement and morphological determination. In *Myzus persicae* increase of mean temperature by 1°C advances the time of progeny migration by two weeks, also well evident in *Acyrthosiphon, Rhopalosiphum, Sitobion*. The impact of temperature on aphid damage and ability of many species to act as vectors as well known. Temperature also contributes to morph determination, besides differences in responses between species, between clones within species and within morphs within a clone. Aphids are very sensitive to warmer conditions with a reduced tendency for alate production, lower migratory potential and delay in sexual morph production. So the direct effect of warming on aphids is likely to be beneficial in relation to individual performance, with important consequences on population dynamics and their pest status.

Impact of elevated CO_2

Aphids being important pests in agriculture and horticulture offer unique opportunities for studies of interactions with climatic changes.

Their rapid and dramatic responses to change in, particular to temperature is aided by their short generation cycles and immense capacity for reproduction. Other variables like CO_2 and pollutants may have a direct or indirect effect on aphid populations. Further, the effects of such climatic changes have their own impact on other host plants, natural enemies, and competitors, making the complexity of their interaction considerable. Life cycle variations for instance, in *Myzus persicae* around the world depend on the interactions between climate and the genotype. The potential of sexual reproduction may be retained throughout the range of the species, but suppressed in some areas due to absence of photoperiodic cues necessary to trigger it, or high temperatures which inhibit photoperiodic responses. Besides photoperiod and temperature, moisture is an important cue for the seasonal behaviour of insects. A behavioural cue hypothesis has been put forward whereby moisture acts as a primary stimulus of crucial seasonal activities in insect life cycles.

Temperature is also a most important factor affecting insect distribution and abundance as in the case of the spotted boll worm *Earias fabia* where the insect life cycle was completed in 88 days when the temperature was 18°C, but at 25°C the total life cycle was 31 days. Rapid changes in the distribution pattern of insects in response to unusual weather patterns are known, several Lepidoptera, Hymenoptera invading new environments.

Agroecosystems when viewed as a whole presents a picture of dynamics of both crop plants, weeds and dynamic components of agroecosystems particularly in view of the microclimatic changes, which become more complex with time. Throughout the growing seasons, the composition of the crop changes in terms of ground cover, volume and maturity characters, so that considerable movement of insect species such as thrips occurs. Equality important are such major pests of agricultural importance as the Brown Plant hopper (BPH) *Nilaparvata lugens* and the plant hopper *Sogatella furcifera* of rice which migrate long distances every year occupying favourable habitats where conditions are suitable, causing considerable damage to the rice crop. Impact of CO_2 on a wide range of phytophagous insects is an aspects which needs more intensive studied and available information points to the low insects abundance to elevated CO_2 levels in respect of insects such as *Empoasca* sp, *Trialeurodes vaporiorum* in different tomato cultivars. It is generally believed that the poorer performance of insects feeding on plants grown in elevated CO_2 is due to the decline in the nutritive quality of their host plants. (Harrington *et al.,* 1995).

Several publications over the last century have dealt essentially with single factors such as temperature, humidity and rainfall, not emphasizing on all climatic variables and to be able to appreciate such changes there is a need for a holistic approach, to explore how whole systems work, since conservation of species will ultimately depend on the conservation of the whole functional system.

Impact of intensification of agriculture

It is well known that crop plants are resilient, containing inherited resistance to attack by pests or pathogens so that sustainability in agricultural production exists in the face of stress factors. All the same, intensification of agriculture has also resulted in the increased production of CO_2, NO_2, ammonia and methane, the ammonia coming also from volatilization and excreta of livestock. One of the greatest sources of global air pollution is the burning of vegetation and forest clearing to make way for agriculture. The use of fire to convert fossil fuels to energy is also elevating atmospheric carbon dioxide concentration which is likely to lead to significant climate change affecting the climate cycle, causing fundamental changes in forest types and distribution. Globally 1 to 2 billion tons of carbon in the form of carbon dioxide is emitted by biomass burning. Acidification of air, soil, water is due to industrial release of sulphur dioxide, nitrous dioxide and carbon dioxide with increased ozone concentration in the lower layer atmosphere or troposphere. Temperature changes and other effects will vary from place to place with increased changes at higher latitudes. An increase of 1-2°C in combination with reduced radiations causes enhanced sterility of spikelets in rice. Of particular relevance are the reduced yields at lower latitudes due to heat and water stress, as against the middle and higher latitudes, where increased carbon dioxide levels will have encouraging results, increasing yields. While industry is largely to blame, intensification of agriculture and live stock husbandry have also increased the rates of gaseous volatiles, methane, and ammonia. Atmospheric methane has increased two and a half times since the beginning of 19th century with natural wetlands and rice fields each producing about 20% of global methane emission, which have their impact on the biota. Annual methane releasing rates at a global level indicate that enteric fermentation, natural wetlands, rice fields, biomass burning, landfills, gas drilling contribute to nearly 95%. Such volatile organics have serious implications in relation to the quality of plant growth, which indirectly affect a host of insects. Sulphur dioxide being an important content of smog can directly damage plants entering through the stomata, interfering with the physiological mechanism of plants and with the life cycle of several insects. All the three pollutants, sulphur dioxide, nitrous oxide and ozone are known to have positive effects on insects of agricultural, horticultural and silvicultural importance, and many of them are prone to rapid, dramatic responses with immense capacity for reproduction. An interesting example relates to the monarch butterfly *Danaus chrysippus* where ozone acts by changing the feeding stimuli of the insect at the plant surface, rather than by changing the nutritional quality of the host plant. Indirect effects are observed in view of the impact of ozone induced effects are observed in view of the impact of ozone induced changes in the free amino acid content of crops. As a result of effective doubling of carbon dioxide concentration projected by 2100, in many trees becoming infested with insects or diseases, which would make them susceptible to wild fires. Besides changes in rainfall patterns have their impact, in particular of greater soil erosion and new crops and varieties will have to adapt to a greater degree of soil and water stress, with greater emphasis on crops improvement involving mixed cropping/intercropping.

Another aspect of anthropogenic sources of toxic elements relate to the contamination of agricultural lands caused by long term use of inorganic pesticides like lead arsenate, calcium arsenate, copper arsenate and mercury compounds used for seed dressing. Increased fertilizer input has its own contribution to soil toxicity, besides several organochlorines used in day to day schedules of crop production. Crops grown on sludge sites have increased metal concentration and all these of serious effects on the soil biota. As a result of combination of changes in the distribution of natural vegetation, changes in natural enemy complexes also follow. In areas of higher temperature, the introduction of new crops of cultivars could have serious implications for existing crops by providing alternate hosts or temporary hosts for many insects.

Temperature and rainfall can influence the efficacy and performance of biocontrol agents like parasites, predators and microbials. The microclimate of many sucking pests will tend to change leading to acceleration of their reproductive cycles, resurgence, behaviour and reproductive potential. Besides, many parasitoids may lose the ``tracking'' ability for the host caterpillars to deposit the eggs due to unsolicited appearance of heavy rainsd. Wasps for instance, may lose the critical ``window period'' for hatching out of cocoons due to late onset of rains. On the contrary, some species of pests like aphids are adversely affected due to fungal outbreaks during the rains. The consequence on the population dynamics may lead to substantial increase in the use of agrochemicals and related load on the environment. Higher temperature leads to faster completion of generation cycle of the pest which indirectly provokes the indiscriminate use of pesticide and hence pesticide resistance.

Pests, diseases and weeds play a major role in agricultural production systems. A chage in 1-2 degree centigrade rise in temperature will have greater repercussion on their biology. Some of the endemic species of weeds that colonize in colder areas may become invasive to warmer parts due to prolific seed production posing a serious threat to the agro-economy. There will be a rapid migration of pests across geographical barriers to warmer climates for faster reproduction and high fecundity. Hence, non-indigenous pests may be yet another cause of concern. The habitat of the monarch butterfly in Mexico described as the `Mecca of the whole insect world' is threatened by excessive logging and incident changes in its environment.

Climate change is expected to bring extension in the host range of many pests and diseases. Instances of sudden pest/disease outbreaks in several agri-horticultural systems are being reported. The interaction between climate and agriculture would mean that any change in climate is likely to result in a spatial right in climatic change, more important environmental factors like soils, water supply, diseases, weeds and insect pests are likely the affected.

On the whole climate change could bring in more complex problems to agriculture. Crop protection will take newer dimensions for tackling these new scenarios. More realistic approach to tackle the changes in the trends of global warming is by good management strategies involving exclusion or early detection and elimination wherever possible.

Climate and Vector borne diseases

Global climate change is likely to hae important consequences on the distribution and intensity of vector borne diseases. Effects of global warming may be more significant than those of man-made environmental chagnes. Change in land use- deforestation, irrigation and urbanization have profound effects on the severity of vector borne diseases. Human transition from a hunter to a farmer was important in the development of human-feeding behavour of several vectors, as also the domenstication of animals, which also became incrasingly abundant hosts. Natural breaks in the forest and traditional vegetational zones being areas where vertebrates tend to congregate and often associated with aquatic habitats, water became an important environmental focus for the contact between vectors that develop in it and the hosts that regularly visit water sources for drinking. As human populations continue to increase and strip the land of its forests and also of the normal blood sources of many blood feeding vectors like mosquitoes and ticks, human-vector contact will tend to increase with increased chances of infection.

One of the most important malarial vectors in India, *Anopheles stephensi* is adapted to breeding in overhead water tanks so that transmission in urban areas is just as intense as in the country side. *Aedes aegtypti,* the dengue vector and the filarial mosquito, *Culex quinquefasciatus* are responsible for transmission of dengue and filarial respectively in urban areas. Dengue is the most important and widespread of human arboviral diseases causing high fatlity rates and is entirely associated with man-mad environments breeding even in flower vases, discarded car tyres, coconut shells, water storage drums and tanks. With ever increasing urbanization and associated human movement, dengue spreads at a alarming rate. The japanese encephalitis virus resulting in brain damage could be fatal, and is transmitted by rice field breeding mosquitoes and it is zoonosis particularly associated with birds and pigs, which act as reservoirs and amplifying hosts. Global warming will cause vector borne diseases like plague and *cutaneous leishmaniasis* which are influenced by temperature. Changes in weather and climate that can affect transmission of vector borne diseases include temperature, rainfall, extreme flooding, draught and sea level rise. Climate related disasters may change the dynamism of human-mosquito contact; floods may create conditions that allow mosquito proliferation and enhance mosquito-human contacts. The influence of *El Nino* on world weather of global warming is well known. The inter-relationship between weather, climate, health is quite intricate and area major cause of climate variability. That vector borne diseases have close connection with climate variation, is equally evident by the eruption of diseases like *cytosporidiosis, leptospirosis* to the warm phase of *El. Nino.* The use of climate forecast information in public health arena becomes important leading to an integrated, interdsciplinary effort involving climate and diseases and epidemiology and such vector borne diseases like plague and cutaneous leishmaniasis are directly influenced by temperature. The use of climate forecast information in public health area becomes important, leading to an integrated, interdisciplinary effort involving climate and disease epidemiology.

By changing entire habitats and communication of insects, temperature changes tend to affect vectors of diseases limiting distribution and local abundance. Temperature controls the time taken for the parasite to mature in a vector, as for instance, JE virus can lead to salivary gland infection in the vector 3-5 days at 32°C and 26°C the same process takes 21 days (Leake and Johnson, 1987). Global warming adds to the alarming rate of spread of dengue. A rise of 2°C could affect the course of intensity of epidemics by lengethening the summer period during which the vector, *Aedes albopictus* is abundant. *Wuchereria bancrofti* which is transmitted by *Culex quinquefasiatus* and with global warming there tends to be an increase in the distribution and intensity of filariasis transmission. Global warming also tends to affect the availability of breeding sites of the mosquito and local vector population tend to peak in the rains (Lines, 1995).

Nearly half of the World's population is infected by vector-borne diseases, resulting in high morbidity and mortality. Though the responsible factors are diverse, a clear trend has emerged : global environmental change has impacted almost every aspect of infectious diseases, from their life cycles to humans' susceptibility to them, fostering their virulence and increased transmission.

Weather affects vector population dynamics and disease transmission, with temperature and humidity considered key variables. Only recently have researchers attempted to predict how climate change such as direct impact (e.g., changes in temperature or rainfall) or indirect impact (e.g., changes in hydrology or agriculture) might affect the distribution of vector-borne diseases with the impact on the distribution and virulence of both the agent and the vector, and resultant vulnerability of the human host. The responses of each element of the disease process to climate change may have ramifications for the others.

The war on vector-borne diseases, especially malaria, was once believed to be all but won. In the late 60's, the period now understood to be the turning point in an all-time greatest global change in the climate, not only did the long-time scourge of malaria disease re-emerged with a vengeance but also new infections like dengue, chikengunya and many others emerged or re-emerged afresh in India and elsewhere in the World! The global environmental change is singularly attributed with the recent epidemiological trends of various vector-borne diseases. Over the past century, global mean temperature has increased by 0.7, which is estimated to go as high 2.5°C by 2050.

It is now understood that warmer temperatures and changing climate conditions impact pathogenic organisms themselves. Heat increases the metabolism of microorganisms, causing more rapid growth and division, and expands the territory they find hospitable. But other effects of climate change upon microorganisms are far more insidious. Heat, as well as increased UV-B radiation due to ozone depletion and pollutants, increase the selective pressure incident upon microorganisms, because they reproduce quickly. Furthermore, these stress conditions can trigger bacterial organisms to express their plasmids. Plasmids are rings of DNA separate from that of the cell; these often code for high-stress coping mechanisms, such as antibiotic resistance and high temperature tolerance. To make matters worse, even bacteria of different species regularly exchange their plastids; as a result, the bacterial genome is much more dynamic and adaptable than that of higher organisms. Thus, a beneficial mutation in one species of bacteria can spread quickly to others, making a mutation-favouring environment, such as that environmental change is effecting, which is a dangerous problem indeed.

Importantly, the most prevalent effects of environmental change upon pathogens will likely manifest themselves in the organisms that transmit the pathogens, the vectors. The prospective higher stress environment affects disease vectors in a manner similar to its effects on pathogens. The ranges of vectors of human and animal diseases are expected to expand as a result of global warming. Furthermore, a higher-stress environment renders pests more dangerous by selecting for tougher strains (Which are often pesticide resistant), and by increasing metabolism. The latter is particularly pronounced in mosquitoes, which transmit several infectious diseases, including malaria and dengue fever. A classical example is the vertical expansion of *Aedes aegypti* in Himalayan ranges upon 1000 ft with an increase of just 1°C. It is understood on the basis of some modeling that a warming of only four degree Fahrenheit (a temperature increase consistent with most estimate of global warming) would more than double the metabolism of mosquitoes, forcing them to feed at least twice as often, consequently increasing the transmission of the infections they harbour.

Environmental degradation due to global environmental change and human activities also forces pests into areas of human habitation that they usually avoid. This phenomenon is particularly evident in rodents, though hardly limited to them. The recent chikungunya fever outbreaks in India during 2006-07 by *Aedes aegypti* and, possibly by *Aedes albopictus*, are the glaring examples of the increased intensity in transmission potential of viruses by these vectors. All of the effects have been exacerbated by societal trends, such as globalization, urbanization and poor health care, but these may not be at the root of the problem.

Indicators of environmental change

It is well known that insect biodiversity is dramatically affected by land use as well as management both in agriculture and forestry and interestingly enough several insects like ground beetles, termites and ants are known to be useful indicators of land-use changes, because the relationship between them and the types of land use depend on the vegetation cover type and on species responses to environmental factors such as soil, moisture and altitude. Ant species such as those of *Solenopsis* can tolerate increased levels of ecosystem disturbance. Qualitative changes in forest arboreal ants such as *Crematogaster* and *Polyrachis* species are also known, not to mention of a general reduction of species diversity of ants with increasing forest disturbance. Every insect species has its own needs such as food, suitable climate, and shelter from disturbance and natural enemies. Termites are good indicators for biogeographical and ecological analyses, termite abundance, biomass and species richness being generally reduced when forests are cleared.

While many macro-invertebrates are utilized for biological monitoring in aquatic environments, chironomids *Kiefferulus barbitarsis* and species of *Cryptochironomus, Polypedium, Tandipes* are known to be indicators of environmental change, in particular of water quality, as also mayflies *Caenis* sp., *and caddis flies such as Macronema* and *Hydropsyche* sp. They are used to define trophic status of lakes, and as indicators of acidification, salinity, and heavy respond quickly to environmental change, as they are able to tolerate low oxygen content due to haemoglobin in their body fluids.

Because of the rapid generation time and mobility of adults, chlronomids are able to respond quickly to environmental change. They tend to be determinate in warm, eutrophic waters and are able to tolerate low O_2 concentration because of te presence of haemoglobin in their body fluids.

Conclusion

A vast range of permutation of climatic variables has to be taken into consideration to be able to predict the distribution, and geographic abundance of a number of species of agricultural, forestry and public health importance. Some species are unlikely to survive climatic changes and this is partly due to the loss of genetic variability. Synergistic effects of climate change with habitat destruction have been predicted to be the major cause of extinction. The loss of keystone species such as pollinators and parasitoids may have critical effects on the whole community. Spatial and temporal changes of host crops tend to have considerable influence on normal enemy complexes and changes in cropping pattern often result in changes in natural enemy complexes. Global warming and increasingly fragmented and degraded landscape tend to result in impediments in relation to insect movement.

In the words of Lawton (2005) ``Global change, be it habitat degradation, climate change or rising atmosphere CO_2 concentrations, simply speeds up the process of changing interspecific interaction in time and space. Initially we observe accelerating changes in the strength and nature of the interaction among a fixed pool of species. Then, with increasing rapidity, the pool itself will start to change, as species become locally extinct and new ones invade, tearing apart existing webs of interaction and creating new ones".

``In longer term, climate change will lead to shifts in the geographic range of species, gradually destroying existing combination of species and creating new ones. Local community richness will tend to decline with these changes and invarious of alien species could compensate for the loss of native species".

Climate therefore directly affects pests, but concomitantly also affects their host plants, competitors, crop, natural enemies and pest management strategies. As climate is an important mechanism for the establishment of success or failure, climatic pattern mapping and climate mapping of a region are important in terms of risk assessment of pest as well as for biocontrol introduction. Ecoclimatic assessment can provide valuable insight in species distribution, in relation to relevant climate data, particularly relating to the assessment of the potential establishment of a particular biocontrol species. In today's scenario, detailed climate data has to be associated with GIS system that provides information on land use and topography, and would help to indicate microclimates of species. As climate matching techniques are well known, which summarize monthly mean, maximum minimum temperature and rainfall patterns, such methods can contribute to pest risk assessment.

Chapter - XI

Overview

A community is composed of all the organisms occupying a site and community composition is determined by identifying spatial and temporal patterns and evaluating factors that determine community composition. The general approach to assess community structure relates to species diversity, species interctions and functional organization. Communities are characterized in terms of the relationships among species, notably feeding interactions involving trophic webs. Insects have been combined into guilds or functional groups based on similarity of responses to envrionmetnal conditions. Identification of patterns in community structures across environmental gradients is an important aspect of population ecology. Patterns are recognized in terms of species diversity, food web structure or guild composition. Species richness usually increases during community development upto a certain point, declining subsequently with the increase in species numbers, the number of species interactions increase resulting in food webs. Fragmentation of habitats often alter species richness. In short, habitat area and stability, resources availability and species interactions are major factors affecting community structure. Species richness generally increases with resource availability and species interactions often affect persistence in a particular habitat. Competition, production, mutalism also tend to affect species directly or indirectly.

As species abundance changes, community structure also changes, altering the network of intractions. Short term variations in community structure reflect interactions among species responding differently to biotic and abiotic condition. Changes in species composition and abundance alter species diversity, food web structure and functional organization. Ecological succession intergrates specific species life cycle strategies, population behaviour, community dynamics and ecosystem process. Insects show successional patterns associated with changes in vegetation, not to mention of heterotrophic succession in decomposing wood and animal caracasses which have contributed in a big way to succesional theory. At the ecosystems level of organization, integration of species interactions and community structure is important. Insects affect virtually all ecosystem properties, especially through their effect on vegetation, detritus and soils and affect primary productivity, being the dominant pathway for energy and nutrient flow in terrestrial ecosystems.

As for community dynamics, differential insect feeding in an ecosystem, affect both the distribution and opportunities for growth of plant species resistant to vibrants of such herbivory. In other words herbivory is a key to ecosystem process that have considerable effects on primary productivity. Mutualistic interactions represent a mutual exploitation of relationships, one member providing a resource that is exploited by the other. Mutualitic interactions often involve modification of host morphology, physiology or behaviour to provide habitats and food reasource for the symbiont. In turn, the symbiont provides necessary resources and protection from competitors or predators. The variety of obligate relationship between pollinators and their floral hosts has contributed with prospects of mutualism. Flower-insect coadaptation in a continuing process, since the dynamics of pollen transfer mechanism are, closely related to the behavioural character of pollinators. Pollinator-flora interactions are essentially a community based one since a constant energy flow through the community becomes. 80% of the pollination by insects are by bees. Besides the social bee fauna *Apis, Trigona* and *Melipona* and the solitary bees belonging to the Apiidae, Megachilidae, Halictidae, Andremidae and Xylocopidae as solitary. Considerable work towards identifying spatial and temporal aspects in community structure and evolving factors that determine community composition. A number of factors determine community composition, distribution and dynamics and the three principal factors describing commonly structure are species diversity, species interactions and functional organization. All these aspects find a place in the chapter on mutualistic interactions and needless to emphasize that the measure of diversity or species richness account of the typical increase in the number of species or number of individuals/species. In short, identification of the pattern of community structure has been a major goal of ecological research. Indices of species diversity, food web structure and functional organization are the principal methods used to compare communities. Functional group organization relates to combination of species on the basis of fundamental responses to environmental variables or effect on ecological process.

Several species of parasitoids associate with a particular insect, in particular localities and such clusters of species contribute a parasitoid community. In a given community we can expect to find various species organized in the form of guilds which are groups of species that exploit the some environmental resource in a similar way (Root, 1967). In short, guilds are often considered as the swelling floods of ecological communities and plays a central role in community ecology. Action which have an impact on one member of a guild might impact other members in a similarly way, such that only one or a few members of the guild need to be monitored. The parasitoid guild associated with the cecidomyid gall midge *Rhopalomyia* includes a torymid, pteromalid, tetrastichid, eulophid, eupelmid and platygastrid. The natural guild becomes restructured when one or more exotic species in the natural guild is associated. In short, the control of host polymorphism depends on the number of species in the guild. The number of natural guilds can be restructured by exotic species. The structure of a parasitoid guild can influence the response of a given member to spatial variation in host diversity. Parasite food webs can contribute to many areas of community ecology, involving the number of parasitoids attacking hosts in different feeding niches and number of parasities that attack the host.

Plant mediated host shifts and competition for a desired niche is well evident in gall thrips. Although galls may be considered to be adaptations to protect gall inducers they can also be viewed as resources for exploitation by a wide range of organisms and there can be no true examples than the *Acacia* gall thrips. Informal, spatial and temporal variations in population and community structure reflect the effect of environmental conditions. Changes in community structure constrain survival and reproduction of associated species. Population density and competitive interactions tend to affect ecosystem function. In view of their capacity to alter ecosystem conditions, studies on insect ecology becomes basic for an understanding of ecosystem structure and functions.

Insect populations are regulated through interactions with natural enemies reflecting essential ecological processes. Habitat stability such as that of cropping systems tends to make the establishment of natural enemies more successful, an aspect equally typical of forest and range lands. Such a stability occurs with the persistence of both the host and parasitoid populations, the factors responsible for such stability being important. Host density dependence, host susceptibility, variable parasitoid sex-ratios are essential factors. Parasites have an innate capacity or limited ability to find prey upto a certain density "mutual influence usually of high parasitoids, with a significant stabilizing effect of low population derivatives of parasitoids and hosts" (Dent, 1993). Community ecology refers to the structure and diversity of parasitoid communities and variability in the number of parasitoid species. Studies on the patterns and processes in parasitoid communities become important from the viewpoint of the alterations in natural ecosystems through human influence. A such, parasitoid community ecology remains a relatively new and expanding field of study - To quote Mill (1994) "Ecological communities such as the parasitoid communities are considered to be structured by a combination of two processes – the colonization of unsaturated niches from a required form of species and the diversification of saturated niches through interactions that involve spared resources or enemies" - Hawkins *et al.* (1992) have investigated the pattern of species richness in parasitoid communities associated with phytophagous insect hosts, wherein the host-larval feeding site tends to affect the colonization of parasitoid species. In short, local parasitoid communities of phytophagous insects represent a colonization of unsaturated niches from a regional species pool (Mill, 1994). The way in which a resource or a particular food in utilized by parasitoid guilds or depend on species which the same resources that are similar in some way that is ecologically relevant. For example, egg parasitoid, larval ecto- and endoparasitoids form distinct guilds - Needless to emphasize that parasitoid communities are to the considered as a set of distinct guilds with increased competitive interactions as evident from their richness in the different guilds.

An exciting field of study, the parasitoid communities tending to become complicated by adding more species, as for example a three species unit commonly with two parasitoids attacking a single host. More than three species are also possible, such as a community of two hosts attacked by three species of parasites – two specialists and one generalist: *Hyblaea peura,* the teak defoliator is a major pest of teak and fifteen species of parasitoids, seven from larvae and pupae. Among the five species which cause recognizable interaction are eulophids, *Elasmus* and *Sympiesis* and a tachinid *Pelexorista.* Maximum parasitism was by the ichneumonids *Eriborus* and *Stictopisthus* sp. By aggregating on certain patches of high host density the parasitoid leave often patches (Nair *et al.,* 1995). Three types of parasitoid are recognized - connectance, semi-quantitation and quantitative (Memmot, 1994).

Connectance web- With only the presence or absence of an interaction between two species of host and parasitoid.

Semi-quantitative - Provides information about relative activities of different species of parasitoids found on each species of host.

Quantitative- relative abundance of different host species of parasitoids so that all hosts, parasites on host communities can be exposed in the same areas. Parasite niche web can contribute to many hosts of parasitoid community ecology involving (1) the number of parasitoids attacking hosts in different feeding niches and the number of parasitoid species not attacking the host (2) number of parasitoids attacking hosts in early and late successional habitats (3) community complexity.

The effects of global climate change on individual insect, population and communities are diverse, with weather patterns, rhythms and interactions, changing. The large scale effects are strongly synergetic with other local effects such as pollution and landscape fragmentation. Not only will wild population and communities change, but also insect diversity with associated crops. Climate change may be positive for flight dependent species like butterflies with regard to such activities as mate location, egg laying and disposal. Climate change further tends to alter hypertrophic interactions such as insect predation, parasites and pathogens. Insects in elevated CO_2 grow more slowly, consume more food, take longer to developed and suffer heavy mortality. As a result, the need for a study on the continued effect of CO_2 drought, temperature and plant nutrients become important. Temperature changes may also affect parasites directly through reducing the maturation line in the vector. Temperature is the climate change variable for which there is considerable evidence of direct effect on insect life histories - growth, development, survival and fecundity. In short, the restructuring of ecological communities could have endless consequences for a wide range of ecological and evolutionary processes, particularly relating to ecosystem functioning and the effects of biotic interaction of species responses to climate charge.

In short, climate is an important aspect of the abundance and distribution of species. For species to survive changing climate, they must adapt to new conditions or shift in distribution. Many insects have large population size and short developmental periods and this phenology, fecundity and survival, and habitat use, can respond rapidly to climate change. Widespread information of species are known to have expanded this distribution in response to climate change is available.

The need for continued research towards identifying spatial and temporal patterns in community structure and evaluating factors that determine competition cannot be over emphasised, more especially species diversity, species interaction and fundamental organization.

SELECTED REFERENCES

Alam, S.M. 1958. Some interesting revelation about the nest of *Polistes hebrocus* (Vespidae), the common yellow wasp of India. *Proc. Zool. Soc. Bengal.* 11, 113-122.

Ananthakrishnan, T.N. 1978. *Thrips galls and gall thrips*. ZSI Technical Monograph.

Ananthakrishnan, T.N. 1984. *Biology of gall insects*. Oxford and IBH, New Delhi, India.

Ananthakrishnan, T.N. 1989. Perspectives in Insect Mycophagy. C.P. Alexander Nemorial lecture I University of Delhi, 17 pp.

Ananthakrishnan, T.N. and K.G. Sivaramakrishnan 2006. *Animal Biodiversity: Patterns and Processes.* Scientific Publishers (India), Jodhpur.

Ananthakrishnan, T.N. and K.G. Sivaramakrishnan 2007. *Ecological Entomology.* Insect Life in odd environments. Scientific Publishers (India), Jodhpur.

Ananthakrishnan, T.N. and Douglas Whitman 2005. Insect Phenotypic plasticity. Science Publishers, Enfield, USA.

Ananthakrishnan, T.N. and A. Raman 1989. Thrips and gall dynamics. Leiden E.J. Brill.

Ananthakrishnan, T.N. and Viswanathan, T.R. 1964. *General Animal Ecology*. Macmillan, New Delhi.

Appannah, S. 1990. Plant-pollinator interactions in Malaysian Forests.

Attenborough, David, 1990. *The Trials of life.* Little Brown Co. Boston Trust : 227-228.

Barbara, P. and Schultz, J.C. 1987. *Insect outbreaks.* 279 pp. Academic Press, London.

Batra, L.R. and Suzanne, W.T. Batra 1966. Fungus- growing termites of Tropical Indica and associated fungi: *J. Kansas Ent. Soc.* 39(4): 725-738.

Beattie, A.J. 1985. *The Evolutionary Ecology of ant-plant nutrition.* Cambridge University Press.

Berryman, A.A. 1986. *Forest Insects: Principles and Practice of Population Management.* Plenum Press, New York.

Berryman, A.A. 1999. *Principles of Population dynamics and their application.* Stanley Thomas, U.K. 243 pp.

Brown, J.H. 1994. On the relationship between abundance and distribution of species. Amer. Nat. 124: 255-279.

Brown, J.H. 1995. *Microecology.* University of Chicago Press, Chicago.

Courtney, S.P. 1984. Evolution of egg clustering behaviour by butterflies and other insects. *Amer. Nat.* 123: 276-281.

Crawley, M.J. 1986. The population biology of invaders. *Phil. Trans. R. Soc. Lond.* B. 314: 711-731.

Crespi, B. and M. Worobey 1998. Comparative Analysis of gall morphology in Australian gall thrips: The evolution of extended phenotypes. *Evolution* 52(6): 1686-1696.

Crespi, B.J. and L.A. Mound 1997. Ecology and evolution of social behaviour among Ausralian gall thrips and this allies. *In: The Evolution of social behaviour in insects and arachnids.* Cambridge University Press.

Dejean and Giberman, M. 2000. A rainforest ant mosaic : the edge effect. *Sociobiology,* 35: 385-401.

Dennis, P. and Fry, G.L.A. 1992. Field margins: can they enhance natural enemy population and general arthropod diversity on farm lands. *Agric. Ecosystem Envir.* 40: 95-115.

Dennis, R.L.H. and T.G. Shreeve 2003. Towards a functional resource based concepts of habitat: a butterfly biology viewpoint: *Oikos,* 102: 417-426.

Dicastri, F., A.J. Hanson and M. Debunche 1990. *Biological invasions in Europe and Medditteranean basins.* Kluwer Academic Publishers, Netherlands.

Drake, J.A. 1989. *Biological invasions:* a global perspective. John Wiley & Sons.

Ehler, L.E. 1994. Parasitoid communities, parasitoid guilds and biological control. *In:* Hawkins and Sheehan (Eds.) *Parasite communities-* Ecology. Oxford Univ. Press, NY, 472-491.

Elton, C.S. 1958. *The Ecology of Invasions by Animals and Plants.* Methuen, London.

Enserink, M. 1999. Biological invasions sweep in. *Science.* 85: 1834-1836.

Feinsinger, P. 1987. Effects of plant species on each others population: is community Structure Influenced? *TREE* 2: 123-126.

Fitzgerald, J.D. 1993. Sociality in Caterpillars. *In Caterpillars.* Chapman & Hall : 372-403.

Gadagkar, R. 1992. Behavioural plasticity, gambling and social evolution. Prof. T.N. Ananthakrishan Foundation Award, 15 pp.

Gadagkar, R. 2007. *The Evolution of Social Wasps.* James H. Hunt, OUP (Foreword).

Gordon, D. 1999. Ants at work: How an insect society is organized. New York Free Press.

Gottparel, R. 1998. Plasticity in life history traits. *Ann. Rev. Ent.* 43: 63-83.

Grover, R.H. and J.J. Burden 1986. *Ecology of Biological Invasions.* 57-66. Cambridge University Press.

Hallman, G.J. and C.P. Schwalbe 2002. Invasive arthropods in Agriculture. Oxford and IBM Publishing Co., New Delhi. 119-137.

Hamilton, W.D. 1964. The evolution of social behaviour. *J. Theor. Biol.* 7: 1-52.

Hanski, I. and Koskela, H. 1977. Niche relation among dung inhabiting beetles. *Occologia* 28: 203-231.

Hawkins, B.A. 1990. Global patterns of parasite assemblage size. *J. Anim. Ecol.* 59: 59-72.

Hawkins, B.A. 1992. Parasitoid-host food webs and donor control. *Oikos*, 65: 159-162.

Hawkins, C.P. 1989. Guilds: The multiple meanings of a concept. *Ann. Rev. Entomol.* 34: 423-51.

Hengeveld, R. 1989. *Ecology of Biological Invasions.* Cambridge University Press.

Herring, M. 1951. *Biology of leaf miners.* Dr. W. Junk Publications, TheHague.

Holdgate, M.W. 1986. Summary and conclusion: Character and Consequence of biolgical invasions. *Phil. Trans. R. Ent. Soc. London.* B. 314: 733-742.

Howarth, F.G. 1991. Environmental impacts of classical biological Control. *Ann. Rev. Ent.* 36: 485-509.

J.H. Hunt 2007. *The Evolution of Social wasps.* Oxford University Press.

Kobayashi, S. 2007. An overview of techniques for the rehabilitation of degraded tropical forests biodiversity conservation. *Curr. Sci.* 93(11): 1596-1603.

Lawrence, W.F. and Peres, C.A. 2006. *Emerging threat to tropical forests.* University of Chicago Press.

Long, D.B. 1953. Effects of population density on larvae of Lepidoptera. *Trans. R. Ent. Soc., London*, 104: 533-585.

Mani, M.S. 1964. *Ecology of plant galls.* W. Junk, The Hague.

Mangan, R.L. and D.S. Moreno 2002. Application of insect behaviour and population ecology to reduce risk of introduction and establishment of exotic fruits flies. *In: Invasive arthopods in Agriculture* ed. Hallman, G.J. and Schwalbe 243-271.

Mathur, R.N. 1960. A list of insect pests of forest plants in India and adjacent countries. Pt. 10. *Insect For. Bull.* 171: 116 pp.

Mills, N.J. 1990. Parasitoid guilds: a comparative analysis of the parasitoid community of Torticids and Weevils: 30-46.

Mooney, H.A. and Drake, J. 1986. *Ecology of Biological Invasions of North America and Hawaii.* Springer Verlag.

Mooney, K.A. and Agarwal, A.A. 2008. Phenotypic plasticity In *Specialisation speciation and Racism;* University of California Press, NY: 43-57.

Mound, L.A. and Morni, D.C. 2001. Domicile consructing phlaeothripine thysanoptera from *Acacia phyllodes* in Australia. *Systematic Ent.* 26: 401-419.

Mound, L.A., B.J. Crespi and B. Kranz 1996. Gall inducing *Thysanoptera* on Acacia phyllodes in Australia. Host Plant relation and key to genera and species. *Insect Taxonomy.* 10: 1171-98.

Murcia, C. 1995. Edge effect in fragmented forests. *Trends in Ecol. Evol.* 10: 58-62.

Nair, K.S.S. 2003. *Pest outbreaks in Tropical Forest Plantations.* Centre for International Forest Research. 74 pp.

Nair, K.S.S., K. Mohandas and V.V. Sudheedrakumar 1995. Biological control of the teak defoliator *Hyblaea peura* using insect parasitoids - Problems and Prospects. *In Biological Control of Social forest and plantation crops pests. Ed.* T.N. Ananthakrishnan, Oxford, HBH, New Delhi, 75-95.

Peters, R.L. and Lovejoy, T.E. 1992. *Global warming and Biological crisis.* Yale University Press, New Haven.

Price, P.W. 1994. Evolution of parasitoid communities in *Parasite community Ecology.* Ed. B.A. Hawkins and W. Sheehan, Oxford University Press: 472-491.

Rajagopal, 1985. Population estimates of seasonal fluctuations of the mound building termite *Odontotermes wallonensis* in South India *Sociobiology,* 11: 67-75.

Ramakrishnan, P.C. 2002. Coping with uncertainties: role of traditional ecological know in socio-economical land management. PINSA 68(3): 235-254.

Raman, A. 2003. Cecidogenic behaviour of some gall inducing thrips, psyllids, coccids and gall midges and morphogenesis of their galls. *Oriental Insects* 37: 359-413.

Raman, A., Carl Schaefer and T.M. Withers 2004. *Biology, Ecology and Evolution of gall - inducing Arthropods.* Vol. 1 & 2. Science Publishers, Enfield, USA.

Renwick J.A.A. and F.S. Chew 1994. Oviposition behaviour in Lepidoptera. *Ann. Rev. Ent.* 39: 377-400.

Romeis, J. and Shanower, T.G. 1996. Arthropod natural enemies of *Helicoverpa armigera* in India: *Biocontrol Science and Technology*. 6: 481-508.

Samways, M.J. 1994. *Insect Conservation Biology.* Chapman and Hall, London.

Shorthouse, J.D. and O. Rohfritsch 1992. *Biology of Insect-induced Galls.* Oxford University Press, New York.

Shukla, P.R., S.K. Sharma, N.H. Ravindranath, Amit Garg and S. Bhattacharya 2003. Climate change in India, Vulnerability, assessment and adaptation. Universities Press, Hyderabad, 462 pp.

Silvieus, S.I., Clement, W.L. and G.D. Wieber 2008. Cophylogeny of figs, pollinators, gallers and parasitoids. *In* K.T. Tilman (Ed.) *Specialisation, specification and Radiation* University of California Press.

Simberloff, D. and Dayan, T. 1991. The guild concept and the structure of ecological communities. *Ann. Rev. Ecol. Syst.* 22: 115-143.

Southwood, T.R.E. and N. Waloff 1978. The components of diversity. In: *Diversity of Insect Faunas.* Symposium Royal Entomological Society, London.

Stamp, N.E. 1980. Egg deposition patterns in butterflies why do some species cluster their eggs than deposit then singly? *Amer. Nat.* 115: 367-380.

Stork, N.E. 1988. Insect Biodiversity: Facts, Fiction and Speculation. *Biol. J. Linn. Soc.* 35: 321-27.

Susan, P. Worner 2002. Predicting invasive potential of exotic insects. in *Invasive Arthropods in Agriculture.* G.J. Hallman and C.P. Schwalber, Oxford and IBH, New Delhi.

Thomas, M.B., Wratten, S.D. and Southern, N.W. 1991. Creation of island habitats in farmland to manipulate population of harmful arthropods, predator density and emigration. *J. Appl. Ecol.* 28: 906-917.

Thompson, J.N. 1994. *The Coevolutionary process.* University of Chicago Press, Chicago.

Tscharntke, T., Tylimakis, J.N., Wade, M.R., Wratten, S.D., Barqtsson, J. and Klein, D. 2007. In: *Insect conservation Botany.* Ed. Steward, A.J.A., T.R. New and O.T. Levis, CAB International, UK/USA: 383-430.

Vermej, G.L. 1996. An agenda for invasion biology. *Biol. Conser.* 78: 3-9.

Vitousek, P.M., D'Antonio, C.M., Loop, L.L. and Westbrook, R. 1996. Biological invasions as global environmental change, *Amer. Sci.* 84: 468-78.